MONTAIGNE

IN

BARN BOOTS

ALSO BY MICHAEL PERRY

BOOKS

Population 485: Meeting Your
Neighbors One Siren at a Time

The Jesus Cow

The Scavengers

Visiting Tom: A Man, a Highway, and
the Road to Roughneck Grace

Coop: A Year of Poultry, Pigs and Parenting

Truck: A Love Story

Off Main Street: Barnstormers, Prophets
& Gatemouth's Gator

Roughneck Grace: Farmer Yoga, Creeping
Codgerism, Apple Golf, and Other Brief
Essays from On and Off the Back Forty

From the Top: Brief Transmissions from Tent Show Radio

Danger, Man Working: Writing from the
Heart, the Gut, and the Poison Ivy Patch

AUDIO

Never Stand Behind a Sneezing Cow

I Got It from the Cows

The Clodhopper Monologues

MUSIC

Headwinded

Tiny Pilot

Bootlegged at the Big Top

MONTAIGNE

⊷ IN ⊷

BARN BOOTS

AN AMATEUR AMBLES
THROUGH PHILOSOPHY

MICHAEL PERRY

HARPER

An Imprint of HarperCollinsPublishers

MONTAIGNE IN BARN BOOTS. Copyright © 2017 by Michael Perry. All rights reserved. Printed in the United States of America. No part of this book may be used or reproduced in any manner whatsoever without written permission except in the case of brief quotations embodied in critical articles and reviews. For information, address HarperCollins Publishers, 195 Broadway, New York, NY 10007.

HarperCollins books may be purchased for educational, business, or sales promotional use. For information, please email the Special Markets Department at SPsales@harpercollins.com.

FIRST EDITION

Designed by Leah Carlson-Stanisic

Library of Congress Cataloging-in-Publication Data has been applied for.

ISBN 978-0-06-223056-0

17 18 19 20 21   LSC   10 9 8 7 6 5 4 3 2 1

FOR THOSE WILLING TO APPROACH CONVERSATION
WITH SOMETHING LESS THAN A STEAMROLLER

The most beautiful lives to my taste are those
which frame themselves to the common model,
the human model, with order but without
miracles and without extravagance.
—Montaigne

CONTENTS

A NOTE ON TRANSLATIONS

When I quote Montaigne I quote his translators. They constitute multitudes. Here and there I will allude to the specific translation. Mostly I will not. A decidedly non-academic catch-all of my sources is included at the end of this book.

MONTAIGNE

IN

BARN BOOTS

INTRODUCTION

I recently tested an electrified pig fencer by grabbing it with my bare hands. As the fibrillations dispersed, I regarded the unit—now pulsing in the grass some fifteen feet distant, where it landed after I spastically flung it—dumbly, then resolved to take the rest of the afternoon off and curl up with the late Michel de Montaigne's essay "Of the Inconsistency of Our Actions."

* * *

If you recognize the name Michel de Montaigne, you have a head start on me. If you don't recognize the name Michel de Montaigne, well, neither did I, at least not until I birthed my first kidney stone. The journey began on a gurney, whereupon I flopped all sweaty and stricken, the stone putting me through the sort of contortions your leading yogis only dream of. I was self-insured, with a catastrophic health plan that allowed me to pay for one ambulance ride, two emergency room visits, and some excellent drugs—all out of my own pocket. It might have been entrepreneurial tenacity in the face of a four-figure deductible, or it might have been the intravenous Toradol, but at some point while prostrate in a penniless Percocet haze, I had a vision that I might convert my agony into cash by writing a personal essay about the experience. I started taking notes, and kept taking them until the moment two weeks later when I peed out what appeared to be the Devil's own gobstopper.

Next I set out to bolster my personal observations by reading several medical journal articles on the subject of renal calculi. A number of authors referenced a sixteenth-century French nobleman—Michel de Montaigne, born in 1533—who had documented his own kidney stone experience. I looked him up. Bought his book. Read the kidney stone excerpts. Then I got drawn into his essay about thumbs. And then I read the one about the sexual danger of coaches (the kind with horses, not the kind with bad shorts and whistles). And another about cannibalism. And the one about marriage. And flatulence. The guy would write about anything.

I discovered it is generally agreed that Michel de Montaigne invented the essay form, for which I thank him, as it allowed me to convert our shared affliction into cash. But after my kidney stone essay was published, I kept reading Montaigne. Fits and starts, bits and pieces, here and there, but I've never really stopped. All thanks to a crippling pain in my flank.

* * *

Last evening near midnight I was called away from the keyboard to help fight a fire. The flames roared and spit sky-high as we pulled hose, the house as good as gone even as it stood. I had been writing on deadline, with a newspaper column due come morning. As my neighbor's home boiled into ash, the clickety-click musings I'd been stitching together evaporated to irrelevance in a flash.

Over the course of my years I have deployed hundreds of thousands of words, primarily in pursuit of next month's house payment. I draw from my heart, but my eye is always on the mortgage. As a writer I can neither claim nor hold the intellectual high ground. Having closely observed great writers and great fires, I know: My keystrokes are short-lived sparks sent flittering into the void, my books a match struck in a monsoon. When I leave this life my most meaningful acts will have been performed not at the writing desk but at the behest of a volunteer fire department pager—and even then I am frequently party to a team presiding over defeat.

Montaigne once wrote, "Good God, My Lady, how I would hate the reputation of being clever at writing but stupid and useless at everything else!" Yeah, buddy. Every time my old snowplow truck won't start, I lift the hood, understand nothing, and long for some fourth dimension where it is possible to replace a bad starter with a nifty simile.

I come from a family of eminently practical people, most of them equipped to perform fundamental functions along the lines of heavy machinery operation, projects involving electricity (on purpose), the design and production of roads and basements and cellphone cases and hay bales, military service, and so on. Whereas I peddle words and three-chord songs.

When blue collar is your background, you never lose the idea that unless you can stack it or stack *with it*, it doesn't count. (A stack of books? Dubious.) Michel de Montaigne once cast "words and language" as "a merchandise so vulgar and vile that the more a man has the less he is probably worth." And yet he spent the last twenty-two years of his life cranking out reams of that very merchandise.

Thank goodness.

Based on our backgrounds, I wouldn't expect to find much in common with Michel de Montaigne. He is permanently deceased in France; I am temporarily alive in Wisconsin.

He was a nobleman born to nobility; I was born to a paper mill worker and a nurse. He was privately tutored in Latin from the age of two and enrolled in the University of Toulouse to study law when he was fourteen; I matriculated as a barn-booted bumpkin who still marks a second-place finish in the sixth-grade spelling bee as an intellectual pinnacle. He served with the military, held several high-level government positions, roamed around Europe, and hung out with the Pope; I am on the volunteer fire department, had a brief run on the student council, went hitchhiking in 1989, and once said hello to Merle Haggard.

And here's where the trail really splits: at the age of thirty-eight, Montaigne retired to the family estate, desiring, as he put it, to spend the rest of his days "in the bosom of the learned virgins."

Gosh, I wish I'd have thought of that.

From then until he kicked the bucket, Montaigne composed essays while ensconced in a castle tower overlooking his vineyards.

I typed that sentence while ensconced in a room above the garage overlooking a defunct pig pen.

Over the course of 107 essays, Montaigne's writing wrassled with humankind's universal biggies—freedom, power, religion, the nature of existence—and he regularly supplemented his observations with quotes from the likes of Homer, Cicero, and Lucretius. But he also left us with his introspections on drunkenness, radishes, hairstyles, impotence, aromatherapy, politics, and kinky sex. In the words of Terence Cave, author of *How to Read Montaigne*, Montaigne wrote with "no evident centre of gravity."

I know the feeling. But this does not excuse me from searching for that "centre." Ta-Nehisi Coates—a man whose writing does not *flitter*—says his only real responsibility as a writer is "to . . . interrogate my own truths." The vibe of that line reminded me of Montaigne's observation that a man wishing to address his own contradictions must "look narrowly into his own bosom." Cross-referencing Coates with Montaigne yields something simple enough: claiming I write simply to pay the rent doesn't excuse me from the duty of writing to be a better person—and not just for my own sake, and certainly not because the world breathlessly awaits, but for specific and powerful reasons including my neighbors and the future of my children. I need to be a better citizen. It is my *duty* to look narrowly into my own bosom and interrogate my own truths. To clarify my character while meandering in the medium of those *essais*, which word translates variously to English as "trials," "soundings," "tests," "experiments," and "attempts."

What is life but an attempt? What is an electric pig

fencer but a metaphor declaring "let us review our progress?"

One of the keys to Montaigne's enduring resonance is his ability to write about himself and yet create—as Sarah Bakewell puts it in *How to Live: Or a Life of Montaigne in One Question and Twenty Attempts at an Answer*—"a mirror in which other people recognize their own humanity." I can sit in a deer stand fifteen feet above the ground in rural Wisconsin reading the 425-year-old words of a dead French nobleman on my smartphone and marvel at all the attributes we share.

And yet (with Montaigne, there is always yet another *yet*), as Terence Cave points out, one must not get carried away with all the resonance, "otherwise we shall simply be constructing a mirror-image of ourselves, an exercise which leads nowhere." I stand here in my barn boots as a fan, but Montaigne must of course be read as an über-privileged European male whose musings take on a different timbre when read from perspectives neither considered nor appreciated in his times. This is neither concession nor negation; it is simply an optically critical fact. When it comes to self-improvement, "having things in common" is hardly a prerequisite and in fact may be an impediment. I have surrendered an embarrassing chunk of my existence to the Internet and social media. But for every super slo-mo puking GIF I've clicked, I have also clicked into insight on whole other worlds, whole other perspectives, and whole other narratives. I am not always comfortable with what I learn. Sometimes you gotta disrupt your own canon.

In the introduction of *How to Read Montaigne*, Terence Cave points out that Montaigne was prone to "caution and provision," manifested in his regular use of the word "perhaps" and the phrase "it seems to me." These sentiments are echoed by the scholar M.A. Screech, who wrote that Montaigne "hedged . . . with caveats and provisos." The German intellectual Hans Magnus Enzensberger saw Montaigne (positively) as "tentative" and "tolerant." And this, above all, is why I like Montaigne: He is a paragon of fair-minded uncertainty, whose most familiar coda is, *I could be wrong*. As a flat-footed rural Midwestern former-fundamentalist-Christian white boy I have pulled some positional and philosophical 180s in my day (and more to come, I trust, and many hope). I have long said that in no case did these changes come about as the result of high-decibel hectoring, public shaming, a bumper sticker, or a snarky hashtag. And yet, absent those discomforts I might never have questioned my position.

What I have learned is, I have so much to learn. Including how to say Montaigne's name. I've heard podcasting academics and National Public Radio hosts pronounce it mon-TEN and mon-TAN-ya (as in Tanya Tucker, for those of us with specific points of cultural reference). On a podcast in which conservative commentator Hugh Hewitt discusses Montaigne, Hewitt favors "Mon-TAN-ya" while his guest goes with, "Mon-TEN," but for the sake of consistency within the podcast, they compromised on "Mon-TAIN," a decision that I assume would have pleased the moderate Montaigne (whom I also assume would be fine with listening to NPR and Hugh Hewitt in a single sitting). Ultimately, I too opt for Mon-TAIN. Issuing from a northern Wisconsin cheesehead, anything else sounds contrived.

I am writing these words on the cusp of a tumultuous American time. There are divisions among us fresh and refreshed. This book is neither up to that nor about that, but it has weighed on my mind in the homestretch. Over the course of Montaigne's life, civil tensions erupted into civil war. Montaigne responded as a citizen by meeting his public duties, including shuttle diplomacy. In private, he wrote essays that for all their meandering always traced threads of kindness, tolerance, and compromise.

Now and then I receive communication from a reader gone all hurt and purple over something I've written, as if I had donned my titanium underpants, declared myself

THE ALL-CAPS INCONTROVERTIBLE KING OF THINKING, and dropped the mic, when in fact I was attempting to round out my mind whilst mumbling around clad in the patchy bathrobe of diffidence. Burke described Montaigne's essays as "catch[ing] himself in the act of thinking, to present the process of thought rather than the conclusions." That is my hope here. To write of Montaigne in terms of exploration rather than declaration. I admit the angle of my appreciation lacks academic rigor, but I believe Montaigne would not object; he shares up-culture and down-culture with equivalent alacrity, operating under the hearty assumption that your appetite for Seneca's interrogatories on courage neither precludes nor prevents your giggling at a good fart joke. You read Montaigne, you feel like you have a friend. And as you gather up the pig fencer, hands still tingling, you anticipate an afternoon in his presence and realize: I may be a reliable idiot but I am not the *original* reliable idiot.

READING LIKE A CHICKEN

LIKE BIRDS WHO FLY ABROAD TO FORAGE FOR GRAIN, AND BRING IT HOME IN THE BEAK, WITHOUT TASTING IT THEMSELVES, TO FEED THEIR YOUNG; SO OUR PEDANTS GO PICKING KNOWLEDGE HERE AND THERE, OUT OF BOOKS, AND HOLD IT AT TONGUE'S END, ONLY TO SPIT IT OUT AND DISTRIBUTE IT ABROAD.

Have you ever dumped a pail of kitchen scraps into a chicken pen? When I did so this morning there was an eruption of flapping, squawking, and pecking. Every chicken for itself, barging and gorging. These are pigs with feathers. But for all their determination, they lack focus; two or three frantic pecks *here* and suddenly the bird is seized with the idea that the really good stuff is over *there* and claws and cackles its way by frenetic zigzag through the others to get to it. Then it spots a bit of

11

stale bread alongside yonder thistle and it's off on another frantic foray.

So it is I read Michel de Montaigne.

I dip in here and there, highlight, underline, copy, and excerpt willy-nilly. It is rarely a structured pursuit. Outside the *Essais* themselves, I happen upon passages by google search, by tweet, by hot link, by recommendations in the comments section, by chance in the used bookstore, or through an offhand remark in a podcast. As my amateur interest in Montaigne has grown, so has my collection of books about or related to Montaigne. But this collection too has grown organically, which is to say, like mold: under less than perfect conditions and in haphazard directions. On the one hand, I unapologetically endorse the random fun of all this tangential discovery. If Montaigne wrote with "no evident centre," I enjoy reading him the very same way. Conversely, I find myself indicted by Montaigne's own words about birds and pedants. In my defense, I'm not sure anyone who has to look up the word *pedant* can be accused of being one, but as far as *picking knowledge here and there*, well, guilty, you bet. And happily so. While hopping helter-skelter through Montaigne's work—snippet by snippet, epigraph by epigraph, any random page yielding something chewy—I am attempting to piece together a means of self-calibration. A fool's procedure, perhaps, but I am building on Montaigne's precedent:

Such foolishness fits my own case marvellously well. Am I for the most part not doing the same when assembling my material? Off I go, rummaging about in books for sayings which please me—not so as to store them up (for I have no storehouses) but so as to carry them back to this book, where they are no more mine than they were in their original place.

Montaigne included over two thousand quotations in the *Essais*. He referred to his constant cribbing as "gather[ing] a posy of other men's flowers," and his professed standard for quotation was simple: "I only say other people in order to better say myself." This is neither theft nor laziness but humility.

One of my translations describes how Montaigne would "strain the writings of Plato." I like the use of the word "strain," as this implies discernment. I also like the image of pouring essays through a colander and trying to draw conclusions from what can be scraped from the screen— rather than "whole sections or pages." This is a form of discretion. On the other hand (as Montaigne, the king of caveats, would want me to say) a strainer catches only the most solid (or obvious) objects. The subtler elements—the broth—is lost. I may be settling for only the fattest noodles, the most conspicuous vegetables, the most glutinous chunks. The "piths and gists," as writer Jim Harrison put it.

The desire to write about Montaigne puts me in heavy traffic on a tricycle. The experts have been weighing in for over four hundred years now, and I am not one of them. I cannot speak to Montaigne in terms of "transalpine humanism," or "epistemological scholasticism," nor can I embroider up anything capable of expressing Montaigne's "inexpugnable inner otherness." (Honestly, if it's that bad just call a plumber.) I welcome the words of librarian and writer Stefanie Hollmichel, who said of reading Montaigne, "while having extra scholarly knowledge adds to the enjoyment, it is not necessary to the enjoyment." Terence Cave says that Montaigne was "preoccupied with philosophical issues which are relevant to practical human experience." I'm not prepping for a doctoral defense, I'm trying to wedge some reflection in amongst the grocery-grubbing. Montaigne himself said that his writing "only nibbled upon the outward crust," and that his "inerudition" disqualified him from presuming to instruct others. Coming from a guy with Montaigne's education and experience this is a wheelbarrow's worth of disingenuous self-deprecation, but the implication is that we amateurs are free to proceed, and for better or worse, it is exactly this sort of intellectual egalitarianism that leads me—armed with a disused nursing degree and unable to remember that the electrified pig fencer is *electrified*—to type up my thoughts on Montaigne. To sweep together the crumbs of the outward crust and study them upon my palm.

* * *

The poet Hannah Brooks-Motl, who wrote an entire book of poetry drawn from the *Essais*, once said that she has no "reading" of Montaigne's work "in the strong academic sense of the word." I hear myself in this and wonder sometimes, when I read Montaigne, or even *about* Montaigne, how my limited powers of discrimination might lead me to erroneous takes. How I might be misled via the prejudices or weakness of the interlocutor at hand. Am I putting my faith in the proper people? In reviews and prefaces of Montaigne's work, or of books about Montaigne, writers and commenters often remark on the superiority of one text over another, laud the grace of this translation over that (usually citing translators by last name only, further adding to the insider-y feel), or cite the perceptiveness of one reviewer in contrast to the ham-headedness of another. Further confounding the issue is Montaigne himself, who amended his essays until his very death, leaving each translator to make personal decisions on how to handle edits and Montaigne's post-mortal marginalia. And here I sit, reading at face value, with a limited capacity for judgment, continually suspicious of my own conclusions, philosophizing at freshman dorm stoner level without the weed. "It is in minds of middling vigour and middling capacity that are born erroneous opinions," wrote Montaigne, and I wonder: What if I crib him out of context?

It would be perfectly felicitous. He did it to others all the time, regularly quoting poets and philosophers out of context to suit his own purpose. In fact he did it so often that Peter Burke called it *characteristic* of his work. If someone reads my interpretation of Montaigne and says, That is not what he meant, I can only reply, *Well, that is where he delivered me.*

* * *

Many writers have cast Montaigne's work in terms of blogging and social media, and given his favorite subject, it would likewise be easy to cast Montaigne as the patron saint of the selfie generation. But Montaigne's writing was a mirror constructed for looking *into* himself. I try to read his essays in the same manner, even—especially—if I read his criticism of another and recognize myself.

In my day-to-day public life (going to the store or post office as distinguished from yapping or singing into a microphone) I am clinically shy to the point of busting out in sheen-sweat and gut cramps. I cherish extended isolation. When I am out and about, I do not want to talk. And yet, when cornered in conversation, I find myself rattling on and on. Friends or strangers, I just can't shut up. My amateur theory is that the logorrhea is symptomatic of solo pursuits. Trucker, farmer, logger, writer; each is a susceptible vocation. In these lines of work (writer napping in his old green chair excepted) there is little time to rest but much time to cogitate, and ultimately, develop capital "T" *Theories*, manifested in anecdote. Alone, we polish them in our mind. Then we get someone's ear:

> *Once you are off, it is hard to cut it short and stop talking. Nothing tells you more about a horse than a pronounced ability to pull up short. I have even known men who can speak pertinently, who want to stop their gallop but who do not know how to do so. While looking for a way of bringing their hoofs together they amble on like sick men, dragging out trivialities.*

There I am, busted by Montaigne. And as I roll past a half-century, he puts me on notice:

> *Old men are particularly vulnerable: they remember the past but forget that they have just told you! I have known several amusing tales become boring in one gentleman's mouth: his own people have had their fill of it a hundred times already.*

The Cotton/Hazlitt translation is even harsher, implying that repetitively loquacious old men are "dangerous company."

I recently joined a group of musicians who had spent the previous week rehearsing with a friend of mine. After we made our introductions, one of the group mentioned they were going to a nearby Italian restaurant for dinner. "You know . . ." I said, but before I could say more, a band member I had never met previously said, "Is this your story about how Draganetti's got its name?"

Whoa. The preemptive cutoff—from a stranger! Clearly she'd been warned.

Our stories are ourselves, and our well-worn stories are our well-worn selves. How then do we avoid leaving the listener—or the reader—well-worn out? Early on in his writing Montaigne was heavy into the Stoics. While Stoicism and small "s" stoicism are not identical, the latter draws on the lineage of the first, so in describing his every thought, emotion, and quirk, Montaigne was performing a contradiction, and according to Screech he was well aware of it:

> *When Montaigne eventually decided to make the* Essays *a book about himself, he was defying one of the basic taboos of all civilized society and one of the great*

interdicts of European culture. Lovers of self, blind to their own faults, were thought to be lynx-eyed for those of their neighbours. Montaigne took pains to show that he was not like that.

Because I want to live as a freelance writer, and a freelance writer must produce copy, I offer things in print I would never offer in conversation, as did Montaigne:

Amusing notion: many things that I would not want to tell anyone, I tell the public; and for my most secret knowledge and thoughts I send my most faithful friends to a bookseller's shop.

Not so long ago I was signing books at a pleasant literary event attended by civilized people. A man placed his book on the table to be signed, then leaned in closely and in a confidential whisper asked, "How is your left testicle?"

I look at him dumbly.

"The *irregularity*?" he said.

"How do you know about that?!?"

"You wrote about it."

"Well, *I shouldn't have*."

Later I found myself onstage grumbling about selfie sticks and how they are the quintessential metaphor of our self-centered society when it occurred to me that I was currently writing my fifth memoir. We are forever riding the prong of our own petard.

* * *

I wish I could drink Montaigne (or any writer, for that matter) in deep draughts, cogitate him at my leisure, and then reconstitute my impressions via some smooth extemporaneous flow. Just lay it out there like a pro. Unfortunately, my cerebral wiring is not up to the task, and so I default to my higgledy-piggledy ways, snipping and sharing as it suits me. Then circling back through, doing it all over again. I am at best a hazy generalist, and Montaigne warned us about people like me.

But: Back in the day, when my idea of literature was largely limited to cowboy books,[1] my friend Frank spent an entire night introducing me to a whole bookshelf's worth of poets. I was galvanized. In short, all that good poetry led me to write a lot of bad poetry. But that good poetry still steeps within my soul, and writing that bad poetry led me to places where people cared about beauty and art and words and

---

[1] I stand by my kidney stone story, although in the course of researching this book I learned that my first encounter with Montaigne likely occurred in my pre-teens while reading Louis L'Amour's *Bendigo Shafter*, in which the main character carries a copy of the *Essais* in his saddlebags.

craft. Reading Montaigne leads me to write about Montaigne, and writing about Montaigne leads me to read even more Montaigne (and smart people on Montaigne).

Among all the chickens randomly ravaging the slop on any given morning, there is always one who locates a prize hunk of glop, nabs it, then darts into the weeds, hoping to choke it down before the other chickens catch on. The tactic is rarely successful, as there are always two or three other birds in hot pursuit, trying to rip the morsel from the first chicken's beak or snatch it should it fall to the ground. But now and then one lucky bird scores and makes a clean escape. And then, safely out of sight, the bird discovers the treasured goodie is too big to swallow. And so it is you will sometimes return to the pen an hour later to find the same chicken trying to gag down a knob of gristle thrice the caliber of its gullet. Unwilling to turn it loose, the bird stands there, blinking in perplexity.

I am that chicken. I read the experts' erudite, multi-layered, cross-referential deconstructions and am left blinking, uncertain how to proceed, but unwilling to give up, hoping if nothing else to absorb some mental nutrition via proximity and osmosis. If my clodhopper curation yields up a strainerful of obvious, it is still a beginning—but only a beginning. I must study those chunks, then give the strainer a bounce and study them again, and see if the change in position has changed my thinking. Then clear the strainer and pour some more. "I think of reading not as the acquisition of static knowledge," says the writer and critic Maria Popova, "but as the active springboard for thinking and dynamic contemplation." The simple "assemblage of existing ideas," she continues, does us little good unless we sit down with them:

> The mindful reflection and expansion upon existing ideas and views, on the other hand, is a wholly different matter—it is the path via which we arrive at more considered opinions of our own, cultivate our critical faculties, and inch closer to truth itself.

How grateful I am for Popova's use of the word *inch*. There is the suggestion that glacial progress is better than none. That I am free to read Montaigne in first gear. To be patient with myself. To worry less about getting Montaigne "right" and instead—in his own words—attempt to equal myself to my thefts. To continue with the reading habits of a piggy chicken: grabbing what I can, when I can, blinking at the big stuff, but not giving up, because (to quote Screech quoting Montaigne quoting Cicero): "Who can shoot all day without striking the target occasionally?"

ROUGHNECK INTERSECTIONALITY

WE ARE ALL OF THE COMMON HERD.

Michel de Montaigne died in 1592 and was thus unable to attend the 2013 Augusta Bean & Bacon Days demolition derby. His loss, because my daughters and I were there, and we had a blast.

The demolition derby was the concluding event of a three-day culture tour that began at the Great River Shakespeare Festival in Winona, Minnesota. The festival was our younger daughter's first exposure to The Bard, and I was pleased to see her alternately transfixed and giggling throughout. It helped that the performance in question was *Twelfth Night*, which contains enough cross-gartered silliness to hold a first-grader's interest.

On day two of this artful ramble we traveled to a farm just down the road and joined a potluck crowd assembled to watch our octogenarian neighbor Tom fire his home-made cannon. The cultural lessons in play on this day included history (Tom's presentation, while informal to the extreme, includes digressions into Napoleonic warfare and expositions on how best to true wooden spoke wheels), organic chemistry (Tom manufactures his own powder using boxelder charcoal), math and physics (the cannon barrel alone weighs over three hundred pounds; shooting at a target two hundred yards away, Tom must make calculations regarding windage and elevation; the sound waves rattle our chicken coop windows nearly two miles distant), and improvisation (rather than cannon balls, Tom launches Dinty Moore beef stew cans packed with concrete).

And then, on day three, the demolition derby, where a bellowing herd of junkyard fugitives smash customized clunkers into one another until just a single surviving vehicle is capable of meaningful linear motion. Meanwhile hundreds of spectators drink, holler, and gorge on deep-fry, separated from the carnage by foam earplugs and a safety barrier of quarter-inch lath. My girls were wide-eyed and joyous in the dust. If you do not want more dullness, Montaigne once said, you must accept a touch of madness.

As we departed the fairgrounds I asked my daughters if they'd had a nice time at both the play and the derby. It was a leading question. I was setting myself up for a chance to deliver a lecture concerning the "pit" in Shakespeare's Globe Theatre, where for a penny, "groundlings" could stand on shucked nut shells and watch art they could otherwise not afford. "Back then," I concluded, "the same people who were at the demolition derby would also have gone to see the *Twelfth Night*." I was hoping my daughters might intuit that cultural consumption can be effected on a sliding scale, including down there where it's greasy. That both couplets and carburetors sing.

They seemed to take the point, and I did not want to overdo it, so I turned the radio on and handed off to Justin Bieber.

* * *

I first encountered the word "intersectionality" while wasting time on Twitter. Retrospectively speaking, I wasn't actually wasting time, since this tangent led in turn to my googling the term and learning that academics deploy it in reference to the ways our social identities overlap—or intersect—in relation to systems of oppression and discrimination. I mention this to be clear: my use of the word begins there but wanders off the academy lawn almost immediately, heading straight for—among other places—the demolition derby, where if I do not use it correctly in the strictest sense, if I have bastardized or bowdlerized it, it nonetheless serves as a point of reconnoiter for fine-tuning my perspective.

I operate from a position of privilege by a multitude of definitions. I have encountered resistance here and there—mostly predicated on presumptions based on class and geography—but the slights are so slight they are worthy of consideration only in the interest of extrapolation. Based on the fact that I live in rural Wisconsin, a Manhattan-based publicist once called me to say she was planning a book release party for my neighbors and me and to that end, wondered where she might rent gingham tablecloths and genuine straw bales. I guess she assumed I'd supply my own overalls and banjo. I almost went for it, because as it happened, my brother possessed a barnful of straw bales, which I am confident he would have made available at his very special book release party rental rate and furthermore cut me in on a sweet 15 percent, but in the end I respectfully declined, without getting into any discussion about how the last thing my neighbors really want to do is sit around on straw bales while the Perry kid reads at them. Here and there, in and out of print, I have been culturally condescended to by certain rare talents, often due to my willingness to write for utilitarian hire, sell books from the back of my van, and crack cow jokes for cash. That I can handle. In fact, it can be good fun when the rare talents are so turgid in their self-assurance they fail to imagine you might know you're being condescended to. One revels in the moment of smiling benignly, a nifty bit of mental jiu-jitsu in which the condescendee condescends to the condescendent.

Some call it "country dumb."

Montaigne said that "you can link the whole of moral philosophy to a lowly private life just as to one made of richer stuff," thus freeing me to wax reflective about competitive car crashing. But this can obscure the fact that Montaigne also believed our crude base selves ought to strive for refinement. That he saw no reason the dude cheering the demolition derby might not also have an interest in existentialism—and might be preferable company to the preeminent existentialist. "I generally find the behavior and conversation of peasants more accordant with the rules of true philosophy than those of philosophers."

What would Michel de Montaigne know of peasants? He was born in a castle purchased by his great-grandfather Ramon, a successful wine and herring merchant. Montaigne's father, Pierre, carried on the family business and further expanded the family fortune by marrying the daughter of another wealthy and politically powerful family, perfect conditions for spawning a spoiled rich kid. But Pierre was a vigorous, intelligent man who became interested in alternative methods of education while serving as a soldier in Italy. Among other things he had young Michel awakened each morning with gentle zither music and brought to fluency in Latin by a live-in tutor before he was allowed to speak his native French. But prior to any of this, Pierre chose poor villagers to preside at Michel's baptism and become his godparents. He then sent the infant out of the castle to live in a small cottage with paupers until he was a toddler.

On the one hand, this was outright slumming; on the other, if it did not imbue Montaigne with empathy (and I believe it did), it certainly led him to lend more than passing consideration to those outside the castle walls. His father's determination that he experience the "meanest and most common way of living" served, writes Montaigne:

> . . . *to make me familiar with the people and the condition of men who most need our assistance; considering that I should rather regard them who extend their arms to me, than those who turn their backs upon me; and for this reason it was that he provided to hold me at the [baptismal] font persons of the meanest fortune, to oblige and attach me to them.*

I do not take this to mean that Montaigne invited the servants in for game night. The very phrasing of his observations indicates he felt superior, and when I say he was seen (in general, and in context) to have been a fair-minded fellow, I mean it and give him credit for his elevated consciousness; but he was also (as the excuse and epithet has it) a man of his times, willing to berate the help "for a glass badly rinsed or a stool left out of place." "I rail at my manservant," he says, in the Donald Frame translation; the Cotton/Hazlitt translation goes with "rattle my man," so perhaps there were more than curses. "I am fairly lavish with raising my hat, especially in summer, and I never receive such a greeting without returning it whatever the social status of the man may be, unless I pay his wages." I can relate to him from afar, but without illusion: raised as I was (a farm kid eligible for government cheese), were he and I contemporaries I'd have probably been bustin' my hump in his vineyard while he sat in his tower nuzzling the learned virgins—and cornering my sister the maid.

In an episode of *The Partially Examined Life* podcast, one of the hosts comments on Montaigne's lovely non-aristocratic air but questions whether he understood how privileged he was. Despite all I wrote above, I believe he did understand. Not perfectly and with great moral qualifications, but his childhood experience lent him a tempering that at the very least caused him to pause for critical consideration of class inequities as an adult. Recalling his experiences as a magistrate, he harbored no illusions about how courts disadvantaged the poor. "They [have] neither the skill nor the money to prove their innocence." When he writes elsewhere, "It is inhuman and unjust to make so much of this accidental privilege of fortune," we are almost certainly hearing echoes from his childhood.

So. Privileged. Yes. But not of what I call the *blithe riche*.

My most recent visit with the *blithe riche* came to pass in the Minneapolis airport. I was waiting for a flight to California. Two young women on the bench adjacent me were looking over a young man's shoulder as he sent a text. "I told her we have to stay at a Super 8!" said the man, and all three collapsed in hoots and cackles. Each of the trio was ferally slim, clad and accessorized as if posed in a high-gloss cologne ad, and as oblivious to the rest of us as if we were buttons on the Naugahyde. While they waited for a reply, the man said, "We should just go out there a day earlier and do Napa, or Sonoma, or something like that." The word "do" fell from his lips like a dead hornet. The phone pinged, their three heads clustered at the screen, and then it was hoots and cackles again. "She *believes* you!" said one of the women. "She's *freaking!*" said the other, and I realized: they'd pranked their friend by pretending they'd been forced to book a room in a chain motel. The same chain, coincidentally, where I was bound to bunk that weekend.

Them folks and me: I don't think we'd travel well together. Let's just say where I come from, when you say Napa we think car parts.

I got no problem with the rich. Know a few of them—two of whom started with dirt-floor diddly-squat. But the best feel no need to let on. And give their thanks in the quiet hour. These are not the *blithe riche*. The *blithe riche* are the ones just confident enough to talk a little louder than the rest of us, a habit formed from speaking in front of the help as if the help had neither ears nor insight. The *blithe riche* exude certitude. Emanate vague impatience at the idea that anything might ever go wrong that zeroes and commas can't solve. Have never had to hold off on taking the sickly baby to the clinic because the money just isn't there. Have never felt a gut-clench while checking the bank balance against the bills. These are those—as Montaigne put it—who "so negligently and incuriously receive their good fortune."

The *blithe riche* revert me to full roughneck. I grew up poor but not deprived. Qualified for that government cheese but wouldn't have starved without it. Had two college-educated parents who chose to live a lean life. Even in my adult years spent living on next to nothing I was exercising an option. But there's something about never quite trusting the floorboards to hold you that gets your back up when someone's in the coffee shop brassily reciting thread counts. Or in the airport stunting on cheap motel rooms. I don't resent the cash, even if it's inherited; I do resent the glib dismissal of privilege. The presupposing prattle. Even the nobleman Montaigne, living off the legacy of family money, had the good grace to write "Others feel the pleasure of content and prosperity; I feel it too, as well as they, but not as it passes and slips by; one should study, taste, and ruminate upon it to render condign thanks to Him who grants it to us."

In a subcategory of *blithe*, I used to quite cheerfully refer to myself as a redneck. After all, I was a country boy, country raised, back of my neck permanently burnt from summers spent in the fields, busting sod and slinging hay bales. "Redneck" was a statement of dermatologic fact. Figuratively, the term as I envisioned it—*intended* it—was affiliated in my mind with pickup trucks and fishing and good people doing honest work without pretension.

I wish it were still so. But over time two things happened. One, the phrase lapsed into irrelevant triteness. Too many country music singers flanked by their fashion consultants and stubble advisors *reassuring* me they were rednecks. Secondly, there was the ever-growing prevalence of perverse pride, in which a term of self-deprecation became more and more a badge of dumbed-down bellicosity. I have my pride. I have my stubborn pride. I have my stubborn *provincial* pride. But when pride blinds me to reflection—and even more importantly, *compassion*, it's time to reconsider.

When it comes to the First Amendment, I am with

the journalist Jeet Heer, who has written (in the form of a tweet), "On free speech, I'm a Millian absolutist, by default. It's a position with many problems, but all the alternatives are worse." (I like that phrase, *It's a position with many problems*. It seems a most Montaignian admission.) I think you ought to be able to say awful things, and awful words. Among other things, it helps one triangulate your position and prepare accordingly. Furthermore, attempts to avoid all offense spawn torturous linguistic contortions. But at some point, we are talking here about garden-variety civility. If I refer to myself these days as a "roughneck," I do so because it's relatively true, it's what my dad used to call us with affection (when he wasn't calling us "knuckleheads"), but above all it is because I think the term "redneck" will thrive just fine without me.

When Harper Lee died, a well-meaning *Esquire* writer tweeted, "Harper Lee taught us the truth about the America in which we were being raised."

I remember thinking, *Yes*.

Then sportswriter, NPR correspondent, and author Howard Bryant retweeted and responded to the message. "They. We knew the score about the America in which we were being raised."

I puzzled on that a second. Then I got it. Howard Bryant is a black man.

They. As in *you, maybe*. As opposed to *We*. *We* who

knew the truth all too well, long before the nice white lady explained it.

There is a thing called parallax. Having to do with the displacement of an object viewed along two different lines of sight. I first learned about it as a youth in the context of buying a scope for my deer rifle. Howard Bryant reminds me it is present in all perspective.

As proof that inattention to intersectionality can fester into trouble on unexpected fronts, have you ever tried to gas up your log splitter using a "non-spill" fuel nozzle?

I deploy those scare quotes with all vehemence.

I own five "non-spill" nozzles. Three different designs. Thanks only to extra credit I pulled a lame B in my one college statistics course, but I am going to declare those five nozzles a representative sample, and based on my experience with that sample (snapped tabs, jammed valves, abysmally slow pour rates supplemented by errant drips, spritzes, and outright hazmatic dam-busters) my brash extrapolation is that the aggregate effect of the "non-spill" nozzle has been to beach a million miniature *Exxon Valdez*es upon America's backyards. These nozzles are also purported to minimize the escape of volatile organic compounds (known in the trade as VOCs). In fact, so efficiently do they trap fumes that you will frequently find your square plastic gas can ballooned into the shape of a red rubber kickball. When you open the valve, that *whoosh* you hear is a flock of VOCs being released into the wild, which, as they say, begs the question.

Here's how you formerly filled your fuel tank:

1. Remove cap from nozzle.
2. Pour fuel into tank.

Here's how it goes with a "non-spill" nozzle:

1. Grunting like a Russian kettlebell instructor, hoist full fuel container to a point higher than the tank and then invert it, as it will not pour until . . .
2. downward pressure is applied against a spring-loaded valve, which will not occur until . . .
3. the full weight of the full container is directed against a plastic tab the size of a gopher tooth, which happens only after . . .
4. the tab catches on the lip of the fuel tank and . . .
5. you simultaneously twist an enigmatically unresponsive plastic locking device (also spring-loaded) while . . .
6. fine-tuning the angle of the approach by thrusting your pelvis at the fuel can in the manner of a spasmodic pole dancer because your hands are both occupied, but . . .
7. it's OK because finally the fuel is flowing—no, wait, it's not OK because . . .

8. the gopher tooth just gave way, the nozzle plunged into the fuel tank, the weight of the full can snapped the nozzle at its base, and now there's raw fuel running everywhere.

One pauses at a time like this to consider the petitioners, bureaucrats, politicians, and engineers in question, then, as the earthworms drown in diesel, conjures Montaigne:

> *These pedants of ours . . . are, of all men, they who most pretend to be useful to mankind, and who alone, of all men, not only do not better and improve that which is committed to them, as a carpenter or a mason would do, but make them much worse, and make us pay them for making them worse, to boot.*

Now and then I perform a humorous monologue that includes a bit on "non-spill" nozzles. There are always plenty of your NPR environmentally friendly types in the audience, and yet the wry laughter of recognition during the show and the comments in the signing line afterward (including surreptitious notes and whispered tips on how to obtain old-school fuel nozzles on the down-low) bolster my sentiments. When your NPR types are trafficking in black market nozzles, you know you got a problem.

The intersectional environmental damage here goes beyond me slopping hydrocarbons. Speaking from the perspective of the folks I perceive as my original people, if the wan mercury-filled curlicue lightbulb debacle tripped the environmental movement at the starting line (I chalk that one up to blatant early oversell), non-spill fuel nozzles wrapped it in duct tape, dragged it off the bike path, and buried it in a shallow grave. "They" took away our smooth tube and replaced it with an evilly ineffective Rube Goldberg contraption comprised of five separate components.[2] Our worst preconceptions regarding "tree huggers" and "the damn guvvermint" have been roundly reinforced. We resent being forced to "save the earth" by a bunch of unidentifiable someones who clearly need to spend more time walking it—and *working* it—at ground level. "They have already deafened you with a long ribble-row of laws, but understand nothing of the case in hand," wrote Montaigne. "They have the theory of all things, let who will put it in practice."

I am currently in possession of an illicit fuel nozzle. It is comprised of a single hollow tube, which I use to transfer all of the fuel to the tank and none to the dirt. It is non-spill because I pour with care—not because I Love Our Mother Earth but because I Paid For The Gas.

---

[2] Five. Not a guess, a fact. Because after using my boots to dissect it on the asphalt driveway, I counted.

In true Montaigne style, let's throw 'er into reverse and back into this from another angle. While in the company of a self-employed, hardworking, and roughneck acquaintance I have known all my life and would trust—*have* trusted, particularly inside burning buildings—with my life, I told the story of the "non-spill" nozzle and how even "NPR-types" agreed with me on that one. If I was expecting a nod or chuckle, I didn't get it.

"Those people are why we've got those things in the first place."

I was taken aback at the turn. "That's not tr—"

"Yes it *is*." Both his tone and the elevation of his eyebrows indicated this was no longer a conversation, and I let it go.

The exchange disturbed me for days. Inordinately so. I felt weak and stupid for surrendering the floor. As a debater I am a bag of cotton candy. My best ripostes are found at the bottom of a twenty-four-hour stew. But above all I was disturbed by the certitude—the *dismissive* certitude—of his "those people" line.

I can do a pretty good riff on the well-earned stereotypes of public radio. But I also know that unlike my conversation partner, I have shaken the hands of thousands of public radio listeners and a not insignificant percentage of those hands are callused. The fact that they laugh at the "non-spill" story indicates that not only do they get the irony, they've lived the irony.

It's the sneer that's such a gut-punch. The idea that we might have no more interest in examining our own hypotheses than those nozzle designers had in field-testing the nozzle with a few farmers. In his piece "12 Fundamentals of Writing 'The Other' (And The Self)," Daniel José Older notes that every character has a relationship to power, and that power plays out in everything from daily annoyances to historical community trauma. When I go to Farm & Fleet to find all the good nozzles removed by mandate, I feel powerless and thus disgruntled. But how self-indulgent is this boutique pique if it does not evolve into some understanding of those truly disenfranchised from power? Intersectionality only works for the good if we're willing to hit the brakes. Pause, and wave the other person through. Recently I observed from the digital sidelines as a gaggle of academics gasped and snarked about a literary festival panel member who dared admit she wasn't familiar with the term "intersectionality." I wanted to ask the academics, Can you tune a carburetor? Quick, without looking, can you *spell* carburetor?

Michel de Montaigne lived in a time when his fellow citizens were killing and torturing each other by the thousands over religion and royalty and a mish-mash of both.

As a public official and behind-the-scenes emissary to all sides, he witnessed firsthand the deadly fallout of factionalism. He understood the attraction of simply riding with the flow of traffic:

> *Yet do I please myself with this, that my opinions have often the honour and good fortune to jump with theirs, and that I go in the same path, though at a very great distance, and can say, "Ah, that is so."*

As a writer, I don't have a lot of opportunity to talk shop, as I spend relatively little time in the company of writers. When I do, it's a pleasure to nod and complain over esoteric references peculiar to the craft—just as the farmers of my youth complained about shear pins and drought. It feels so good to find someone who responds with, "Ah, that is so." But Montaigne knew that "Ah, that is so" will only go so far, and that is where I am culpable. I am weak of character in that I sometimes nod—"Ah, that is so"—when I do not agree at all. There is cowardice in this, but also a deep strain of Scandinavian reserve. M.A. Screech has observed that Montaigne's "natural complexion" was to resist "all vehemence," and:

> *He refused to use the distorting language of hate even about religious opponents. This would lead one to ex-*

pect Montaigne to have been in favour of religious tol-
erance. No doubt he was, but mainly from necessity.

During a recent election the majority of my hometown county voted for a presidential candidate I regarded as a gold-plated tinhorn. That portion of the electorate includes more than a few of my friends, neighbors, and relatives. I do not agree with their position, but I am not surprised by it. There was a time back in the pre-podcast 1990s when I was paid just under seven dollars an hour to listen to a fellow on the radio each day, then deliver a synopsis of the show for one of his fans. I can say with neither snark nor irony that I learned a lot—about society, about the fellow (who was, as the poet Seamus Heaney put it, a "mouth athlete"), and about myself.

What Mouth Athlete taught me is that if you spend too much time nodding in vehement agreement, you lose track of what's going on to your left and right. Thus the same man who wrote of reveling in the head-nodding "Ah, that is so" moment, also wrote, "Harmony is a wholly tedious quality in conversation." But Montaigne didn't stick the word *conversation* in there by accident. He assumed we would convene in disagreement, but that we would indeed convene, and once convened would converse without intermediation. By surrendering our independent thought to "interlocutors" (be they talk show hosts, pundits, blog blasters, or Top Commenters) we bypass our neighbors, and thus their humanity. We allow the mouth athletes on their distant electronic thrones to cast epithets at the folks next door. We are excused to mock the movement without giving consideration to the people. To spit criticism without context. I recently sat in a fishing boat on a remote Canadian lake with a man who spends half his life driving a log truck and the other half living on generator electricity in a wilderness cabin. We had a lovely time fishing and shooting the breeze. The subject meandered to child-rearing, and when I mentioned that we had done some homeschooling, he perked up and congratulated me on prying our kids from the grip of "lefty thugs" and their "agenda." When I explained that our children were currently enrolled in the local public school system and we were having a positive experience, he spent the next five minutes explaining to me that we were not. Later, when I stopped by his cabin, the generator powered lights were on and the mouth athletes were going full blast, and I understood: listen long enough to the man yelling from a distance and you come to trust him more than the man in the boat beside you.

I despair in these situations because I *liked* the guy in the boat. I *related* to the guy in the boat. I was *raised* by folks like the guy in the boat. But who am I if I let this sort of talk go unchallenged? The best I could muster in this instance was a mumbling demurral that was instantly steamrolled.

I too want to be certain. I want to nod. We in the muddling middle are often embarrassed by our noodle tendencies. And I do need to be stronger. Montaigne took stands, and took them in spite of personal danger. But he also understood a solid stance is not synonymous with solid thinking. I believe he would have nodded approvingly at this passage, written by Seamus Heaney's former editor, Craig Raine:

> It is no accident that when Heaney first began to write, he signed himself Incertus. It expresses an existential truth about his moral configuration, his helpless, deliberate and conscious commitment to awkward complication.

Here's an awkward complication: I believe some criminals need killing (and in certain nightmare scenarios I would happily do it myself) but am opposed to capital punishment. This is less a position of principle than it is a position of admitting I would be entrusting just application of the death penalty to the same government that produced a blurting non-spill fuel nozzle.

* * *

Aptly enough, I was on Twitter when I read the following Montaigne quote:

> *There is really no greater or more persistent folly, nor anything more anomalous, than to be excited and annoyed by the world's fatuities.*

—which pretty much encapsulates an average ten minutes scrolling the medium. In a review of the book *Reclaiming Conversation*, Jonathan Franzen wrote, "Our rapturous submission to digital technology has led to an atrophying of human capacities like empathy and self-reflection, and the time has come to reassert ourselves, behave like adults, and put technology in its place." First of all: As the father of a smartphoned teen I know a thing or two about reasserting myself and putting technology in its place. Second, and more germane to our discussion, I disagree with that first clause. The Internet—and social media in particular—is the ultimate intersectionality catalyst. Voices previously unheard are amplified. "Social media made a new level of visibility possible for a diverse group of writers previously shut out of legacy[3] publishing," said Alexander Chee in a tweet I only saw because it was retweeted by

---

[3] Loosely (and not unanimously) defined as a handful of big publishing houses in New York City.

the aforementioned Daniel José Older, a New York–based Cuban-American writer I encountered on Twitter while procrastinating on a project for a legacy publisher. We find ourselves interwoven in unexpected ways. In less poetic terms, whether you shoot someone's favorite lion or jack up the price of a lifesaving drug, you can't hide like you used to, and I'm not naïve about the hateful, deceitful, cesspools festering with nation-poisoning spawn—heaven forfend you be a woman who disturbs the trolls. Raised in the hands of the masses, smartphones are the new pitchforks, and every fortress is porous.

Of course I agree with Franzen that we must not succumb to an all-digital world—just ask my teenager. Our negotiations are ongoing and dramatic and the floor joists supporting the hallway leading to her bedroom may not survive the stomping. But it is also important to note that the Twitter account[4] that generated the Montaigne quote broadcasts him nonstop, 140 characters at a crack, right in there amongst all the fluff, outrage, and dangerous lies.

Set the proper filters, and technology enables the benevolent infiltration of fresh thinking. In my case, I especially cherish the opportunity for the roughneck to intersect with the intellectual. Montaigne said he would prefer his son learn to talk in a tavern, mocked universities as "yap-

---

[4] @TheDailyTry.

shops," and regularly elevated peasants at the expense of pedants ("I prefer the company of peasants because they have not been educated sufficiently to reason incorrectly") but this is breezily specious: sans academe and academicians he would have been just another trust fund loudmouth; despite his protestations, time after time he drives home his points by quoting philosophers. Just as his memories of the pauper life caused him to question his privilege, his exposure to philosophers drove him to question his own thinking. Finally, taken at face value the above quote implies all commoners are good company (as long as they don't overthink things) and university professors can't change their own motor oil. Gimme ten minutes, three phone calls, and one stop in the alley of a bar called The Joynt and I can muster a lineup to the contrary.

Among Montaigne's many gifts was an ability to translate both ways between all stations. If you've spent much time reading academic journals and papers, you'll assume this gift is not widespread. In fact, this has become one of my more well-ridden hobby horses, and I recently penned up notes toward an essay on the subject, including the line, *How can we be shocked at the rise of proud anti-intellectualism even as the academy engages in terminal auto-obfuscation?* Shortly thereafter, I encountered a Nicholas Kristof column decrying the incomprehensible jargon of academic publishing and felt further validated. I was reading Kristof and nodding in agreement, (*Ah, that is so . . .*) when, as happens in this hyperlinked age, I wandered off into a world where his conclusions were being questioned. This tangent led me to the blog of Dr. Tressie McMillan Cottom, PhD, where I shortly discovered that in addition to working as an assistant professor at Virginia Commonwealth University, she wrote for a number of non-academic publications. That she worked hard to make esoteric but important academic information understandable and accessible.

One of her most effective tools? That dreaded civilization-rotter, Twitter.

I put my hobby horse out to graze.

Over the next month or so via Dr. McMillan Cottom's Twitter feed I learned that she—and academics in general—face headwinds from all directions: from within their own institutions, which often retreat at the first hint of controversy and leave their faculty stranded; from the fact that established faculty want to *remain* established, and so it is often the hungrier—and more expendable—adjunct faculty who embark on these public forays; and, finally, roaring anti-intellectualism expressed through blunt force digital antagonism: Dr. McMillan Cottom says the most commonly expressed negative sentiment in the comments section of her blog is "some form of 'who the fuck do you think you are,'" which reminded me of Christopher Hitchens writing what it was "to have spent so long learning so relatively little, and then to be menaced in every aspect of my life by people who already know everything, and have all the information they need."

Academia tapes targets to its own ass. But using the most egregious examples to make the whole school a fool is a dumb[5] trick. "Our mind is strengthened by contact with vigorous and orderly intellects, and debased by constant intercourse with mean and feeble ones," wrote Montaigne, which is to say the onus is on *us*, to seek out thinkers that challenge and edify *us*. (Note his use of the qualifier "constant" in the previous quote, meaning if we scroll through a few happy puppy memes to get there, so be it.) When someone suggested I follow Daniel José Older on Twitter simply because we were both writers with backgrounds in emergency medicine, I wound up reading his books and articles and clicking on his YouTube video about cultural italicization. Because someone in my timeline retweeted Jeet Heer, I now look forward to reading linked Twitter essays composed by a Canadian Fulbright scholar with a sense of humor. Because I googled "academese" to confirm my preconceptions, I discovered a sociologist who blew them apart.

---

[5] And powerfully effective.

Few things online have broadened my mind in a manner commensurate to books and travel—but there is a static formality to books that can lull us into forgetting the human at the other end exists in an unedited state, and most travel involves other formalities that keep us outside certain perimeters. When Daniel José Older Tweets a photograph of his possibly malevolent cat, it reinforces my sense of Older as a real person *alive right now* rather than just a name beneath a title. When Howard Bryant lobs a frustrated "@VW" Tweet, I learn we are both afflicted with fraudulent Volkswagens. Through Dr. McMillan Cottom's Twitter feed I discover we both cherish Dolly Parton, boiled dinner in cold weather, Robert Duvall in *The Apostle*, the poems of Lucille Clifton, and Lawry's Seasoned Salt. And neither of us has seen *Star Wars*. By side effect, I am more receptive to her academic considerations on the study of race as an axis of stratification (which she kindly renders in terms I can understand). The day she described handling a hater by acting "country dumb," I busted into a giant grin.

This is pixilation rendering us as *more* human.

"In our conversation with others, silence and modesty are most useful qualities," wrote Montaigne, and the more I remember this the more I am helped. My insights on race relations are granular and anecdotal, accumulated through chance encounters, a handful of intimate personal relationships, and family photographs that assume a more multitudinous hue with each passing year. But none of this puts me at the epicenter of expertise. None of this anoints me to "weigh in" or insert the ubiquitous and infamous "Well, *actually* . . ." Sometimes the finest thing I can do is shut up and redirect. Amplify other voices. Let them speak so, as Montaigne wrote, I might be spared "the vice of confining [my] belief to [my] own capacity." As I write this book, there is much talk—both derisive and defensive—of "trigger warnings." If you want my opinion on the matter, I won't give it; I will direct you to read Dr. Tressie McMillan Cottom and essayist Roxane Gay on the subject, as they are operating from an informed first-person position. Experience, rather than talk-show fueled ridicule. There is enough secondhand 'splaining in the world.

Sometimes the best we can do is give each other breathing room. Roxane Gay has questioned the assumptions that are made about us if, in the wake of certain events, we do not make the standard social media declarations (manifested as slogans, hashtags, and avatar adjustments):

> *Demands for solidarity can quickly turn into demands for groupthink, making it difficult to express nuance. It puts the terms of our understanding of the situation in black and white—you are either with us or against*

us—instead of allowing people to mourn and be angry while also being sympathetic to complexities that are being overlooked.

This "demand for response" as Gay puts it, is also ferociously impatient. You must tell us how you feel about this *now*. To which Gay responds:

The older (and hopefully wiser) I get, the more I want to pause. I want to take the time to think through how I feel and why I feel. I don't want to feign expertise on matters I know nothing about for the purpose of offering someone else my immediate reaction for their consumption.

This is utterly Montaignian. It is also why I so hungrily want to converse with people with whom I disagree or even lack enough information to hold an opinion. I want to ask questions and not be assigned a "side." I simply wish to avoid succumbing, as Jean Starobinski wrote in *Montaigne in Motion*, to "the idle fictions which gullible men allow themselves to believe." Idle fictions are comfortable fictions; in order to escape them, I must put myself in a position to be discomfited. Expose myself to what Terence Cave refers to as "the tension of difference," currently available via handheld screen.

*   *   *

This year I attended three hometown funerals in a single week. The pews and visitation halls ran heavy to rough people living tough lives. Lots of cowboy boots and concealed carry.[6] In stance and palaver, I felt utterly at home. We never completely shed the comfort of our cohort. The comfort of knowing the code. And yet I am learning—and subscribing to—new codes. New intersectionalities that press a wedge between me and the comforts of the past. So many guns. So many big bad pickup trucks. So much By God bravado. So much fear.

In *Montaigne and Melancholy*, M.A. Screech points out that Montaigne lived and wrote during a time when "inherited certainties" were under assault:

> *—assaults from classical literature made widely available; from Church Fathers edited and translated; from original texts of the Scriptures, translated, glossed and preached upon; from rival schools and methodologies within every university discipline, not least philosophy; from New Worlds discovered and from the Old Worlds of China and Japan, with all the troubling impact of*

---

[6] For the record, I don't wear cowboy boots.

their venerable cultures based on premises other than
those of Jerusalem, Athens, and Rome.

Well, hello Twitter cross fire. Cocoon if you wish, but click around a bit, and soon, as Screech puts it, "the exemplum is stripped of its preeminence, its privileged permanence." I might be bending his intent, but I believe what we're talking about here is the inability to *avoid* viewing conflicting views. "An understanding of history destroys innocence," said the historian John Hope Franklin. Among the people at the funeral, it is common to thank God for giving us this great country. I fly the flag in my yard and second the gratitude for all this flawed experiment has provided so many of us, but at the very crumb-like least that prayer should be offered while mentally genuflecting beneath the handmade sign at Wounded Knee.[7] It is too easy, as Christopher Hitchens wrote in his Thomas Jefferson biography, to forget that when America had less than five million inhabitants, nearly one-fifth of them were African slaves. Some got bootstraps, some got the boot heel. But even that allegory won't do. It is too

_____

[7] "Once you really know the history of Native Americans, if you put a sheet over the American continent you can see all the places where the blood soaks through." —Jim Harrison, from *Conversations with Jim Harrison*.

easy to cast these things in the abstract. Thus Ta-Nehisi Coates maintains his tenacious focus on slavery's physical, pain-riddled destruction of the individual. After reading Coates, the next time I sat through a prayer thanking God for "giving" "us" this great country I opened my eyes to look at my own small daughter and focused on what it was to see one's child die in a frozen gulch beneath the bark of a Hotchkiss gun. This mental exercise repaired nothing, but it has changed the way I talk.

None of this requires me to grovel. I'm not being cowed or browbeaten into some utter cultural capitulation. If I have the phrase right, it's *Check* your privilege, not abandon it. It's not really much of an ask. To disrupt the canon is not to destroy it but rather recast it. To—this word again—*amplify* it. I want every day to remember that mine may not be the definitive perspective. That this country may have arisen from something other than bald eagles, gumption, and pickup truck commercials. You can't plant a seed in blood and expect it will flower up in peace.

I retain a knee-jerk defensiveness on behalf of my roughneck class, just as I do for the fundamentalist Christians who raised me. After all, they have frequently and fundamentally proved their loyalty to me. But I want to be a better person. To the people around me. *All* the people. And so I'm hacking away at it. Fear of conflict, fear of saying the wrong thing, both hold me back more than they should. I find it helps to reread this passage from Daniel José Older's "Writing 'The Other'" piece:

> . . . *some weird, ultimately meaningless mind game about irony, ego, and self-perception—one I was playing all by myself—was keeping me from doing work I knew was important. The same is true for writing about other cultures and experiences. You will jack it up. You'll probably jack it up epically. I know I have. This doesn't mean don't do it. It means challenge yourself to do it better and better every time, to learn from your mistakes instead of letting them cower you into a defensive crouch.*

In a Twitter discussion on this topic, Ta-Nehisi Coates echoed Older: "Saying stupid shit is part of exploring other worlds . . . getting embarrassed is part of learning. You don't get to learn and look cool." Often I wind up wedged, as Montaigne once so vividly put it, with my arse between two stools.

But: No whining, and don't just sit there. Montaigne piled up a lot of words, but he supplemented them with action, serving his community and kingdom. And now we're back to why I still carry a fire department pager. Not exactly pioneering for social justice (and these days I'm down to a handful of calls per year) or pulling clandestine meetings with Henry of Navarre, but bridging some gap, taking some fundamental action, and as ever experiencing more compassion-engendering intersectionality. Yesterday I logged two first responder calls. Both were for frail individuals struggling under precarious physical and financial circumstance. In each instance the winter wind cut straight through the blankets we tucked around the cot. When I hopped back inside my warm van after the second call, some well-fed politician was on the radio proposing we cure the nation's ills with a dose of get-up-and-go, and it occurred to me that few phrases are more smugly presumptive than, *Get a job!*

* * *

Sometimes I wish I was shinier and smoother and cooler and groovier and had what they call *panache*. I wish I was smarter, and could hold great minds in rapt attention. Take the academy by storm. Win all the high-end prizes.[8] Then I find myself standing at a podium telling stories in a high school gym in the tiny northwestern Wisconsin town of Luck, and midway through the one about my neighbor Tom and his beloved wife Arlene, I am reminded for the fifty-seventh time how fortunate I am. Just before I spoke, a woman walked in and to no one in general, said, "Who's

---

[8] Someone did once build a giant plywood replica of one of my books and pulled it as a float in the Cheese Days parade, so deep-fry that and eat it, Pulitzer committee.

got a Ford pickup with a mail carrier sticker on it?" and when six people said the same name at once, the woman said, "Well, tell 'er her dome light is on." I know better than to paint naive caricatures of quaint small towns and bucolic rural life (that said, it is simply truth that someone slipped a plastic-covered paper plate of homemade cookies into one of my book boxes). Nor is this about "common folk" united and living in gosh-darnit harmony. It's about being able to talk about books while wearing camouflaged hunting boots because they work good on the ice.

That night my hosts provided me a motel room south of town. Outside the window I could see snow, a tag alder swamp, and an empty pulp rig parked on the backside of a truck stop. On the wall above my bed were two Robert Mapplethorpe prints. Somehow I have been allowed a life in which I can look at my camo boots beside the bed and that log truck in the lot and give thanks for my rough-neck roots, even as I give thanks for the artists and poets who taught me to spot a Mapplethorpe. I hope I am not portraying the pursuit of a pan-cultural palate as a self-congratulatory act. I am thinking in terms of gratitude. In my brief Shakespeare speech following the demolition derby I did not lay upon my daughters some onus for becoming "well-rounded." Rather, I said, what you have done here is expanded your opportunities for joy in this life.

Sometime after my trip to Luck, I spoke at a grade

school in Illinois. I was met at the door by an administrator who informed me that many of the students were children of Mexican laborers and spoke English as a second language. I got to use some of my clunky Spanish, picked up from my bilingual wife and my Panamanian and Ecuadorean relatives.

When I give these talks, especially in rural areas, I try to bust up that whole "writer as precious flower" thing and instead focus on my working-class background as a kid raised on a dairy farm. During the Q&A, a boy raised his hand, and in far better English as a Second Language than my Spanish as Same, said, "You told us you grew up on a dairy farm."

"Yep," I said, all farmer-y, all working-class shitkicker.

"Did you milk your own cows?"

It took a split second for the import of his question to smack me upside the preconception. Here was your intersectionality, in real time. Your intersectionality, and your parallax. Where I come from—or when I come from—we who owned cows were the strugglers who came to the Christmas concert late and smelling funny in hand-me-downs. From his perspective, my people own the cows, but his people do the milking.

"Yep," I said. "You bet." And he smiled at me with approval. I felt my parallax shift, and refreshed my resolve to pitch in with those facing forward rather than those fighting backward.

CONFOUND THE FOOL

WERE IT NOT THE SIGN OF A FOOL TO TALK TO ONE'S SELF, THERE WOULD HARDLY BE A DAY OR HOUR WHEREIN I MIGHT NOT BE HEARD TO GRUMBLE AND MUTTER TO MYSELF AND AGAINST MYSELF, "CONFOUND THE FOOL!"

This morning when I fed the chickens, I noted their water was drawn low, and so set to replenish it by opening the valve on our rainwater reservoir. This homemade "system" collects runoff from the granary roof and is rigged together with reclaimed and repurposed items, from the secondhand rain gutters to the industrial poly tank that used to contain I-know-not-what but has so far not caused tumorous drumsticks or glowing yolks. Due to the scavenged design of the system—combined with the fact that it is completely gravity-powered—I always feel environmentally pious when the water gurgles out the hose. This feeling evaporates—goes up in steam and curses, in fact—thirty minutes later when I leap from my desk upon realizing that I never closed the valve. Sprinting across the yard I arrive to find the tank drained. Specifics of the subsequent soliloquy shall remain a secret between me and the feathered critters, all ruffled up and clucking at my outburst like church ladies who just caught the pastor peeing in the punch.

Here is the exponent to my anger, the thing that speed-ratchets me past crestfallen and straight to incandescent, the item that truly launches my rage-rocket: in that bucolic moment when I was topping off the feeders and reflecting upon the earth-loving gurgle, I made a mental note to my-self, *Don't forget to shut it off.*

I tell myself this because . . . *I am forever forgetting to shut it off.*

And then, in the six seconds it took me to return the feed pail to the bin, my mind wandered, my feet followed, and I went tripping la-di-da back to the writing room, leaving the tank to exsanguinate. Three hundred arti-sanally curated gallons, currently percolating down-valley where the neighbors will run it through their sump pump.

And I did the exact same thing last week.

Are you chuckling?

Why not. I am too.

Sorta.

Sorta, because these incidents make for funny anecdotes in time, but in the moment my rage is real. Lava-hot and hateful.

Profane.

"You fucking *dolt*!" I roared.

At myself.

Out loud.

That's not funny.

Montaigne's admission that he wished to grumble and mutter at himself for being a fool is found in his essay "That We Laugh and Cry for the Same Thing." In another translation, Montaigne seems angrier, using the phrase "growling to myself," and referring to himself as "You silly shit!" The essay itself focuses more on contradictions in our character and behavior than it does absentmindedness, and I therefore deploy it somewhat out of context, but by Montaigne's own example we are forever free to depart the central theme and pursue all tangents. "Of one subject we make a thousand, and, multiplying and subdividing, fall back into [the] infinity of atoms," he wrote, and regularly cast himself as a scatterbrain[9] subject to the "disorderly

---

[9] Bakewell quotes a Spanish ambassador of the time describing Montaigne as "a man of understanding, though somewhat addle-pated."

sallies" of a mind "rushing [or roving] about here and there." Terence Cave says this "erratic, unpredictable flow of thought" is reflected in Montaigne's writing, where "the apparently rambling syntax mirrors the movement of the mind." Montaigne happily described his writing as "un-stitched," and left it up to us to follow:

It is the undiligent reader who loses my subject not I. In a corner somewhere you can always find a word or two on my topic, adequate despite being squeezed in tight. I change subject violently and chaotically. My pen and my mind both go a-roaming.

What a glory it is, if you are of an undisciplined mind, to let it off the leash. Free it to pursue all digressions. Montaigne once characterized his as a runaway horse. Mine is a nervous squirrel trapped in a spinning bingo tumbler filled with acorn-scented Superballs. And yet the same capricious synapses that drive me (and those near me) bonkers when all I want to do is remember to shut off the chicken water become a switch-backed roller coaster of delight when the only goal is discovery. What fun, in the early draft of a thing, to let the mind meander as Montaigne's did, at "the gait of poetry, all jumps and tumblings."[10]

---

[10] Peter Burke writes that Montaigne admitted he was inclined to

Sometimes at author conferences I am asked to describe my writing "process." First, I giggle. Then I suggest that I'm essentially operating via "functional ADD." I intend the latter reference lightly and in a humorous vein—as the parent of a child who has struggled mightily with same, I am well aware of the distress it causes. (During the diagnosis session, my wife, Anneliese, kept looking at me and raising an eyebrow as if ticking off my behaviors box by box.)

I did the "functional ADD" bit many times without issue. Then I shared it at a conference where—unbeknownst to me—a clinical psychologist was in attendance. The following day, she sent me an email chastising me for using the term loosely. After listening to you talk, she wrote, it is my professional opinion that you do NOT have ADD; the symptoms you describe are consistent with *flight of ideas*, which is associated with a completely different *manic-depressive disorder*.

Well, whatever it takes, I thought. Then (as a self-employed writer who had been buying his own health insurance since 1992): *Hey! Diagnosed for free!*

In *Montaigne in Motion*, Starobinski writes of the essayist's "natural disposition," observing that it "never settles

---

imitate Seneca, whose "relatively loose and informal" style was known as the "Senecan Amble."

into any stable equilibrium: it manifests itself, rather in a series of opposed states." Then he quotes Montaigne:

> *Now I am ready to do anything, now to do nothing . . .*
> *A thousand unconsidered and accidental impulses arise*
> *in me. Either the melancholic humor grips me, or the*
> *choleric; and at this moment sadness predominates in*
> *me by its own private authority, at that moment good*
> *cheer.*

Manic-depressive? I don't know. I understand the term is outmoded, replaced with bipolar. I can't speak for Montaigne, but to cast my peaks and valleys as clinical, when so many encounter far worse, would be self-aggrandizing. I get in the pit now and then, but I hesitate to write of my depression, not out of shame but because while it has often been incapacitating, it has never drawn near deadly. But I do know what it is to sit strapped in lassitude, chest leaden with inertia, vinegary weakness leeching the guts, draining all initiative as the spirit curdles into a self-loathing fetal hunker.

But then, *shazam*! The manic phase arrives; sweet, giddy relief. I allow myself to get caught up in music, I type rough drafts in a rush, I stride out to our ridge overlooking the valley and breathe deeply, exultantly. I sprint up the hill to my office when only hours ago I slogged. I am well aware that I'm riding currents of chemistry but when the cold river reverses, you swim the warm flow like mad; when the clouds clear, you run the meadow at the gait of poetry, all jumps and tumblings.

*   *   *

Grandma always said I was a tad drifty, and I've been proving her right all my life. There was that time in college I pulled the front wheels off my Plymouth Duster to work on the brakes and Mom called me to lunch just as I was finishing, after which I proved that you can make it through two stop signs, three right-hand turns, over five miles and up to fifty-six miles an hour before the left front wheel pops off because the lug nuts are back there in a pile on the shop floor. More recently I wrote 482 words of a column eulogizing my old manual typewriter before nagging déjà vu drove me to scroll up and discover I had already written that exact column three weeks previously, landing me right in the middle of a Montaigne quote: "I, moreover, fear . . . the treachery of my memory, lest, by inadvertence, it should make me write the same thing twice."

In Montaigne's time, memory and intelligence were seen as one, and the phrase "he has no memory" meant "he is stupid." "When I complain that my memory is defective [people] either correct me or disbelieve me, as though I were accusing myself of being daft," says Montaigne, referring to every person who ever said, "We just talked about that!" "We went over this yesterday!" Or, "Why don't you just write yourself a reminder?"[11] Trouble is, many of these failures occur within time frames and situations in which writing one's self a note is impractical. In the category of driving off with things on my car roof I count one wallet (circled back and found it in the Culver's drive-through; celebrated my good fortune with a second order of curds), one iPhone (heard it thump the luggage rack, then watched in the rearview mirror as it pinwheeled down the highway), and an infinity of coffees, each of which I placed on the roof "just for a second" only to forget it in two.

One of the great joys of being self-employed in a room over the garage is that when I need to take a leak, I just hop off my treadmill desk (sadly, not a metaphor, but rather a defense against arthritis and doughnuts), place my reading glasses on the back of the chair beside the door, step outside, and avail myself of nature. And then, invariably, I re-enter, remount the treadmill, get it up to speed, and realize I forgot to grab my glasses from the back of the chair. Here is Montaigne prefiguring me:

---

[11] I often do . . . and then get triply pissed when I overlook the reminder.

There is nobody less suited than I am to start talking about memory. I can hardly find a trace of it in myself; I doubt if there is any other memory in the world as grotesquely faulty as mine is!

Based on a calculus including the number of days I work from home per annum, the hours I spend at my desk per day, and the amount of coffee I drink, I forgot to grab my glasses 1,590 times this year alone.

In my day-to-day interactions with the world, I comport myself with reliable restraint. In fact, I have been told by those who love me that I am overly laid-back in this respect and get run over in the process, but I have no appetite for fighting. Even when cut off in traffic I opt for the sardonic aside over cursing and flying the bird.[12] During blood pressure–busting customer service phone calls I choose dispassion over harangue. But when the idiot is within, I am merciless. When I back out of the parking stall and the coffee cascades down the windshield, I curse myself. When I dismount the treadmill to fetch my glasses I mutter self-abuse all the way.

---

[12] This streak recently shattered on Highway 13 south of Bayfield when that pinhead came at me across the double yellow. For which I bestowed upon him double birds. While steering with my knees.

"The most terrible and violent of our affections is to despise our own beings," wrote Montaigne, and my hottest rage is reserved for the greatest fool I know: Myself. Which makes me wonder: Who's yelling at who? How can I step outside of myself to rebuke myself? When I'm yelling at myself is myself also cowering in the face of this rage? Or is my first-self feeling the great cleansing rush of *finally*—after resisting the urge to do unto others—allowing myself to unload on the idiot in the instant? When I worked in a mental health unit many years ago, I was trained to note when a patient was prone to rapid changes in emotion. The term we used was "labile." I wonder how many times I wrote that in someone's chart before I forgot it, only to recall it again recently while berating myself solo.

Who are the two people in my head? Because there are clearly two people. One saying, *Don't put your coffee on the roof,* the other saying, *Gonna anyway. Despite all prior experience.* In the foreword to Screech's *Melancholy,* Marc Fumaroli speaks of "Montaigne's *moy,* which is at once the director of his conscience and the directed." Why does my *moy* reject *moy*'s good advice, when no one knows *moy* like *moy*? It is as if I have access to a person who knows more about this one thing than anyone else in the world and still I roll the dice.

*   *   *

Lack of retention is one of the chief frustrations of my life. I get things when I see them. When I watch my father rewire a fuse box, it makes perfect sense. When I try to do the same thing ten minutes later, it is as if someone switched the colors on all the cables. Definitions escape me the moment I close the dictionary. The joke I made a few chapters back about looking up the word *pedant* was no joke. There are words—*pedant* is one, *pedagogy*, *humanism*, *heuristic*, *phlegmatic*, *tautology*, and a multitude of others included—which I can never seem to retain with any clarity. Rather I hold them in a fuzzy impression, able to assume their meaning in context but having to look them up again and again. It is worse with conceptual doo-dads like Stoicism, or Epicureanism. I have a good sense of what they mean, but cannot recite the definition. I can do a two-hour humorous monologue before a theater of strangers on three minutes' notice, but tell me a "Nun walks into a bar . . ." joke on Tuesday, and I'll enjoy it just as freshly Wednesday. When I read a novel for the second time, the characters are only a shade less unfamiliar than the first time. And the plot? Forget it. How grateful I am, then, to be reading Montaigne's essay "On Books," and come across these words: ". . . if I am a man of some reading, I am a man of no retention."

My *man*.

Absent-mindedness and poor retention are not the same thing, but they are definitely first cousins, and in my case, dating. I got good grades all through high school and college—even testing out of many prerequisites—but by and large survived by cramming,[13] intuition, extra credit, and essay questions.[14] Perhaps my trouble now is I no longer have the patience for memorization. Back in my community theater days I memorized great swathes of dialogue; now the only memorizations I commit are of stories for my own monologues or lyrics to my own songs—and even these aren't really memorized: If you ask me to recite the lyrics of one of my songs starting from the second verse, you might as well ask an emu. Then again these rely less on memory than momentum. Other mind-blocks are even stranger: If I think of Rickie Lee Jones, I am immediately reminded of—but can't recall the name of—Joni Mitchell, and vice versa. Same with Kurt Russell and Jeff

---

[13] This tendency continues post-graduation: When we were awaiting the birth of our daughter I could do ten minutes on Braxton-Hicks contractions—today, I recall nothing except they're sorta fake.

[14] Everyone else hated 'em but I loved 'em. Especially in nursing school. I'd just write and write, one line after the other, until time was up. I've since had visions of my instructors plowing about three-quarters of the way through, and then, overcome with the sense that they were drowning in a bullshit riptide, scribbling an "A" and moving on.

Bridges. And sometimes Kirk Douglas and Lloyd Bridges. It is as if each resides down an adjacent hallway, and once you visit one you can't visit the other.

And yet, I have a photographic memory for useless details: The frost-fuzzed roofing nails visible on the underside of a haymow roof in January; Grandma's Teaberry gum stuck to the plasticine of her cigarette box; the ringlet of hair just below the earlobe of that girl at the gospel service. And hotel rooms—I have near-perfect spatial recall of hotel rooms going back over a decade of book tours. I can tell you right where the desk sat in relation to the refrigerator in the New Orleans Hampton Inn & Suites; I can fully reconstitute the layout of that Motel 6 just outside St. Louis; I can tell you there were louvered curtains in that fancy Kansas City hotel where they hand-delivered an apologetic note about the ladybugs.

But useful retention? The struggle is never over. And so I read and reread. Get the gist over time, but in the evanescent manner of Montaigne:

> *I eternally fill, and it as constantly runs out; something of which drops upon this paper, but little or nothing stays with me.*

I am lost without access to the original source. As such, I have never been much of a quoter or a reciter. I can do the whole of the childhood favorite "Animal Fair" (the drunk monkey version), the *Hee Haw* staples "Gloom, Despair, and Agony on Me" and "PFFT! You Was Gone!" and I can muster up the final stanza of Dylan Thomas's "Do not go gentle into that good night," but nothing more—just enough so I can rattle it off at band sound check and project a bogus literacy. When friends start tossing out favorite movie lines, say, quoting bits from *Monty Python and the Holy Grail*, I'm the doofus grinning and nodding, and saying, "Yeah! Yeah! The rabbit one! Do the rabbit one! Pointy teeth! Pointy teeth!"

You will understand then, my relief (and creeping sense of superiority) when I read the following quote in a piece about Montaigne in the *London Review of Books*:

> *People who forget readily and rapidly can see things freshly and reason back from what they deeply know, rather than having to rely on received opinion.*

Well, sure. Let's go with that.

* * *

The term "absentminded" conjures images of a loveable fuzzyhead. Anneliese can set you straight. It is a grind to live with someone who asks the same questions over and over, who goes out to sort the recycling and winds up at the library, who can't abide an extraneous apostrophe but dreams blithely past the screen door that has leaned against the siding for weeks. Apart from the frustration, Anneliese is ten years younger than I, and I worry sometimes that she is just biding her time, gathering the evidence necessary to place me under professional supervision. One day when I was loading book boxes and musical gear into my van (and running late as usual), I kept reminding myself not to forget the cash box, then took off and made it forty yards up the driveway before I realized I forgot the cash box. Slammed 'er into park, ran back for the cash box, then jumped into the car and shot out the driveway only to find my way blocked by the van I'd left idling less than sixty seconds previous. I snuck a peek at the kitchen window. Anneliese was watching, and possibly taking notes. Buster, I thought, one of these days she's gonna make the call.

*   *   *

My favorite book about Montaigne is a petite red hardcover missing its dust jacket. I bought it online, used, for twenty bucks and change. The title was daunting: *Montaigne's Discovery of Man: The Humanization of a Humanist*, by Donald M. Frame. My initial intuitive affection for this edition may have been triggered by its petite dimensions, subliminally suggesting it was digestible for even a plodder like me. I am subject to fits of intellectual optimism.

The thing is, once I got going, that book really did treat me fruitfully. It was the first work that helped me put Montaigne in context across the arc of his philosophical development. In short, Frame maintains the essayist began as a Stoic "advocating rigid self-mastery and a firm struggle for consistency against the ills of life," shifted into a "skeptical crisis" summarized by the slogan "What do I know?" and ultimately entered an Epicurean period emphasizing moderation, "simple human goodness," and a balance between serving oneself and others. I have since read others who would fine-tune or quibble, but it was good enough for me, working as I do at Sesame Street level. It was also a necessary reset: I had fallen into the habit of largely highlighting the Stoic bits of Montaigne—in short, only the bits where my inner voice said, *That's me!*

Step by tiny step, *Montaigne's Discovery of Man* eased me beyond this. It *was* digestible—as long as I took tiny bites. I'd sit with it for a few pages now and then, pen and pink highlighter in hand so in the instant of revelation I could translate Frame's erudition into my own pidgin. As the book was of tuckable dimensions, I carried it on book tours, on family trips, to the local coffee shop, into the bedroom and out to the deer stand. From San Francisco to New York City, from Duluth to the Caribbean, it was my traveling companion for the better part of a year.

I had seven pages to go when I boarded the flight to San

Diego. The rest of the book was covered in graffiti like a boxcar parked too long outside the train yard: scribbles, circles, underlines, arrows, and rambling marginalia, all against a backdrop of pink highlighter swipes. Precious insights, critical observations, life-altering wisdom—all destined to enrich the very book you hold in your hands. As soon as we were allowed to lower our tray tables I got to work on those remaining seven pages, and when I completed my final notation there was still an hour to fly. Standing to go to the restroom, I placed the book in the seat pocket.

Don't put it in there. It's an actual voice I hear.

I put it in there.

And lo, later in the hotel room, there were lamentations long and lugubrious, gnashing of teeth, and rending of the empty backpack.

* * *

At the age of thirty-seven, Montaigne was out riding when a stampeding horse plowed into him. He was thrown a great distance, knocked unconscious, suffered facial contusions, vomited blood, lost his vision, and was carried off for dead. He would later use the accident as a focal point for rumination, and a number of scholars have deconstructed the incident at length, but I'm most interested in a 2002 *Psychiatric Times* article in which Dr. Michael

Sperber proposes that Montaigne suffered a traumatic brain injury (TBI), defined as "a non-degenerative, non-congenital insult to the brain from an external mechanical force, possibly leading to permanent or temporary impairment of cognitive, physical, and psychosocial functions, with an associated diminished or altered state of consciousness," and that this may be the source of his faulty memory and rambling writing style.[15]

Some crafty fact checker may wreck my fun on this, but to the best of my recollection (he says, in this the chapter certifying his faulty recollection) I was on the field of play for all but four seconds of the 1982 New Auburn Trojans football campaign, as I started both ways—offense and defense—and played on all special teams. (The four-second break came when the coach let someone else in for one play.) I wish this record might stand as testament to my exceptional athleticism, but the truth is we played eleven-man ball and only suited up seventeen, several of whom were shrimpy underclassmen—unlike monstrous me, having through the throwing of haybales burly'd myself up to 165 pounds plus peach fuzz.

So I played a lot, and loved it. And loved it unapolo-

---

[15] Terence Cave's aforementioned observations on Montaigne's rambling syntax as a reflection of his mind were drawn on the very passage in which Montaigne describes the horse mishap.

getically for the violence. I was a mostly well-behaved and mostly nonconfrontational kid (a few schoolyard scuffles but hardly Jets and Sharks), and the football field provided my one chance to tee off and wallop someone with the full approval of all involved. I didn't play dirty, but I happily lowered the boom at every opportunity (unless we were playing those lumberjacks from Birchwood, in which case I did a lot of illegal holding and tried to conversate my way out of contact, all the while wondering if I'd ever be able to grow a beard like that).

I say I lowered the boom. Sometimes the boom dropped on me. Six years I played, and more than once found myself blinking and shaking my head in the wake of a collision. I once leaped for a pass in practice and met Clay North helmet-to-helmet with such a crack that within my skull lightning flashed and I could feel the distinct cerebral divisions of the longitudinal fissure and central sulci. To this day the only image I retain is of our hands reaching for a football silhouetted in a blaze of sunlight.

There were other non-football brain-bouncers, mostly collected over the normal course of being a kid in the country. Later, during my bicycle racing years, I twice crashed headfirst with sufficient force to shatter my helmet, and was once left with headaches and a tangy postnasal drip. Then there was the bull that knocked me silly in Wyoming. And the time I woke up bleeding from the scalp at the base of an oil rig in Wyoming. And the double-knockout "draw" boxing match in Wyoming. And the forgetting to duck while running under the overhang of that cabin in Wyoming.

Wyoming: The Concussion State!

When not raging at myself over my own forgetfulness, I have earned a significant portion of my living by poking fun at it. It is an evergreen resource. But when I read that researchers at the Mayo Clinic have found one out of three adult men who participated in amateur contact sports in their youth demonstrate significant brain damage (and that's without serving time in Wyoming) I turn back to the Sperber piece, which contains this passage:

> *[Montaigne] said his memory was hopeless. If a thought occurred to him that he wanted to write down, he had to tell someone else at once in case he forgot it before he could walk to the room where he kept his paper.*

I carry a notebook in my back pocket.[16] If a thought occurs and I don't write it down in the instant, it is lost.

---

[16] The intervention is not foolproof. Yesterday I discovered a Post-it note in my pants, upon which I had scribbled, "*Lectured about curlicue. No. All I have to do is turn it on.*" I have NO IDEA what this means.

Am I just wired this way? Or are those football collisions reverberating after all these years? Did I lose more than temporary consciousness when I took that header off the oil rig? Grandma's quote about me being a tad drifty predates all these episodes, but still, I wonder. And my mood swings? The flashes of anger? Are they residual of all the cranial clobbering? Am I proceeding at the gait of poetry, or just tumbling?

My younger daughter and I are driving to meet my parents, with whom she will be staying overnight. She is jabbering at me from her booster seat, some stemwinder that ends with both of us laughing. I make a mental note to appropriate the anecdote for a newspaper column due later that day. After I drop her off I realize I can't remember the story. I can visualize the stretch of road we were driving as she spun it—we were at a stoplight on the overpass when she hit the punch line—but none of the details. I have this *sense* of it, a foggy impression of it floating just beyond reach, and sealed within a smoked glass orb. There is this idea that a pinprick recollection will pierce the glass and clear the smoke. I pull to the shoulder and call my parents. They put my daughter on the phone. "Remember that story you were telling me in the van when we were driving through Altoona?"

"Yes."

"I can't remember what it was about. Can you tell it to me again?"

She does. I thank her. "I hope you have a nice time with Grandma and Grandpa."

"I will, Daddy," she says. I close the phone and reach for a pen.

And I can't recall the story. *None* of it. Nothing but the image of us at that stoplight. My gut goes cold. It isn't funny, it isn't frustrating, it's frightening. I sit there for two full minutes, straining to penetrate the smoky orb. Finally I redial.

I am caught off guard by what it feels like to be embarrassed to explain myself to a third-grader. She giggles and tells me the story again. This time I scribble notes as she speaks.

"OK, Snort, I got it now," I say, with a lightness I do not feel. I am about to say good-bye, when she speaks.

"You can call me whenever you need to remember again, Daddy."

My heart cracks, and like light, love spills out.

SHAME

. . . THE BETTER I KNOW MYSELF, THE MORE DOES MY OWN DEFORMITY ASTONISH ME.

Now and then someone will approach me, usually at a literary event, and make a nervous inquiry. "I was looking for your books online, and wondered . . . did you write a book called . . . called . . ."

And then they trail off.

"*Hand Job*?" I ask. Verifying, mind you, not offering.

As the person cautiously nods, I explain that the book is a catalogue of hand-drawn typography compiled by a graphic designer with whom I share the same name, although he's better at catchy titles.[17]

---

[17] I once wrote a book about (among other things) building a chicken coop. Put chickens *and* the coop on the cover. Titled it *Coop*. Seemed perfect, until the very first radio interview of the

At this point the inquirer giggles with relief. I assume the nature of the reaction is based on my previous oeuvre; no one expects me to write about sex.

Least of all me.

I'm getting sweaty just thinking about it.

And not in the hot-to-trot way.

But I'm gonna have a whack at it.

I could have worded that better.

"It's strange to think at how young an age I became a subject of sex," wrote Montaigne in his essay "On Experience." "My lot is like that of Quartilla[18] who could not remember being a virgin."

The delayed-adolescent dude in me is dying to raise one eyebrow, shoot the man an elbow and a wink, and lay a chortling "you and me both, pal" on that quote. But thanks to Mom and Jesus, I was real late outta the gate. From my teens I liked girls and making out with girls, but even well out of college was still not doing The Deed

_____

book tour, when live on the air the host said, "Tell us about your new book . . . *Co-op!*"

[18] A character from the Satyricon, a Latin work of fiction written in the late first century AD and one of those things you read when you need to be reminded that sexual hoo-hah did not begin in the 1960s. Or, no matter what you've seen on the Internet, someone's always been doing it.

because I had been raised in a religion decreeing that I should only ever have sex with one person, and that person should be my antecedently obtained and legally ordained heterosexual spouse. Although I had begun drifting from the church and spending time at second base and sometimes getting caught in a pickle between second and third, I still nonetheless held the prize and power of Heaven in my heart and retained a residual worry that this misbehavior would land me on a greased pipe to Hell.

There was also a certain prideful stubbornness in play; you hold out for so long you're disinclined to cave.

So much buildup.

As it were.

In a book I read in a cabin somewhere, Harold Bloom said Montaigne took a pragmatic view of sex. Later, I saw Alain de Botton on public television suggesting this pragmatism was perhaps a product of Montaigne having spent much of his life in an agrarian setting, where he saw animals doing what animals do, and thus did not operate under the illusion that human humping was any more sacrosanct than hog humping.[19] In fact Montaigne refers to this "most confused [and common] of our actions" as God's way of putting us all on equal footing with "the fools and the wise . . . and the beasts."

---

[19] To be clear: Alain de Botton did not use the phrase "hog humping."

To her eternal credit, my mother was very matter-of-fact about sex, providing me with all the details by third or fourth grade. This is of critical note, because A) she is a devout Christian who holds sex as a sacrament of marriage, and B) I have yet—in this my fifty-first year—to hear her utter a single double-entendre or off-color comment. The time she told me I was not conceived in Hawaii (as I had long believed) but rather during a three-day ice storm in Illinois, was a stunning one-off, and simply a matter of setting the record straight.

So you might have expected Mom to play the speak-no-evil prude. Far from it. Rather than treat sex as a squirmy secret, she laid it right out there—in terms of physiology, sure, but also as a most delightful gift. "You can even do it standing up," she said once, and if the frankness widened my eyes then, it makes me smile today. In fact, she had witnessed relatives suffer as a result of functional and fundamental gynecologic ignorance, and resolved it would never be so with her children. I'm grateful for that. She even provided us access to an Anatomy & Physiology textbook with colored art. I loved to draw as a child and spent hours copying out skulls and brains, but also drew and re-drew the genitourinary systems of both sexes. It had to be tough for Mom, but as long as I did my drawing downstairs and in the open she let me be (she was not so forgiving when she was delivering clothes to my closet and discovered a horrific pen-and-ink drawing I did of dismembered women— this could have been grounds for therapy were it not for the fact that I was an otherwise stable grade schooler and my pal Vinnie Boscoe and I had lately been reading up on Ed Gein[20]).

Basically Mom filled in the blanks I hadn't filled myself by watching our farm animals in their natural procreative pursuits. (Here my experience echoes Montaigne's—although the fullness of his experience did not include watching the artificial inseminator lean shoulder-deep into the back end of a cow, a semen-charged straw clenched in his teeth.) Thanks to Mom's forthright instruction, I didn't have to wonder, or rely solely on my pal Marco and his brother's *Playboy* collection for instruction. And during one particular pre-date discussion on the porch steps that I recall as clearly as if it occurred this afternoon, she very possibly saved me from becoming a teenaged father by about one overheated hour (there was this new neighbor girl . . .).

---

[20] A murderer and body-snatcher from Plainfield, Wisconsin, near where I was born. My parents frequented a hardware store owned by one of his victims.

Mom, you did a good job. You have endured so much. You saved me from so much.

Mom, you might want to stop reading now.

Sarah Bakewell has written that Montaigne "wanted to know how to live a good life—meaning a correct or honorable life, but also a fully human, satisfying, flourishing one." It's that "but also" clause that informs this chapter; therein lies a rich fraction of existence. My ingrained sexual reticence has carried into not writing about it, but Montaigne would have found my attitude quaint and uptight: "Our life is part madness, part wisdom. Whoever writes about it merely respectfully and by rule leaves out more than half of it . . . one must waive the common rules of modesty for the sake of truth and liberty."

Oh, he waived them. Whether describing postcoital habits of the ancients ("They wiped their fundaments . . . with perfumed wool") or a persnickety penis ("The indocile liberty of this member is very remarkable, so importunately unruly in its tumidity and impatience, when we do not require it, and so unseasonably disobedient, when we stand most in need of it") his ruminations regarding "all commerce with Venus" are consistently frank, celebratory, and pragmatic. From the Frame translation:

What has the sexual act, so natural, so necessary, and so just, done to mankind, for us not to dare talk about it without shame and for us to exclude it from serious and decent conversations? We boldly pronounce the words "kill," "rob," "betray"; and this one we do not dare pronounce except between our teeth.

Montaigne then suggests that the less we talk about it, the more we think about it: "Does this mean that the less we breathe of [sex] in words, the more we have the right to swell our thoughts with it?" And why is talking or thinking about it a bad thing?

[Sex is] the most noble, useful, and pleasant of all [Nature's] operations . . . on the other hand she lets us accuse and shun it as shameless and indecent, blush at it, and recommend abstinence. Are we not brutes to call brutish the operation that makes us?

Montaigne does make one wry concession: If the sex act results in the creation of "such a stupid production as man," perhaps it is appropriate "to call the action shameful, and shameful the parts that are used for it." But ultimately, Montaigne casts fanatical modesty as "hypocritical prudishness" and equates it to treachery manifested in contradiction:

Convention forbids us to express in words things that are natural. And we obey it. Reason forbids us to do what is wicked. And we ignore it.

If I am to write about life, implies Montaigne—if I am to write about *Montaigne*—I must write about sex.

Of course Montaigne doesn't have to face my mother this Sunday.

And at the age of fifty-one, I still care what my mother thinks.

I shudder to consider the state I'd be in if I didn't.

So, Mom, apologies. Gonna write about s-e-x. Apologies, also, to those readers who have supported me (and, ergo, my family), book after book. Who send me kind notes, leave generous comments, and share encouraging words in person. Not a single one of which has said, *Hey, we'd love to get your thoughts on doing the nasty!*

To them, I offer a mea culpa à la Montaigne:

My apology is addressed to those of certain kinds of temperament (who are I believe numerically greater than those siding with me). I would like to please everyone, even though it is a difficult thing [as Cicero said] "for one single man to conform to so great a variation in manners, speech and intentions."[21]

But that's the thing: You never know who's reading. Sometimes the insights that save us—or at least ease our journey—come from souls we've never met. Somewhere between my mom and Montaigne, perhaps I can walk the fine line between simply snickering or oversharing and—even if by accident, or by virtue of my own lame example—help someone else navigate these complications.

An apology, also, to those odd few expecting the deep nitty-gritty. By the standards of the day, what follows will be a PG-rated meander. No triple-X prurience. No names, and limited specifics. In *Montaigne in Motion*, Jean Starobinski said Montaigne's book "cross[ed] the threshold of the boudoir" into "forbidden territory," but Montaigne also had some cover: his privileged position (no Twitter clap-back for a sixteenth-century nobleman; no former partners dishing on Facebook) and his sense of mortality: "I speak the truth not so much as I would, but as much as I dare, and I dare a little the more as I grow older." Based on life expectancies of the time, he assumed death was

---

[21] The quotation within this excerpt is Cicero translated from the Latin.

sneaking up the sidewalk, and this led him to pledge, "For my part, I shall take care, if I can, that my death discover nothing that my life has not first and openly declared." I assume no breath other than the one already drawn, but by the actuarial standards of State Farm I've got a few more decades ahead and need bridges more than ashes, so I am not willing to pledge the same.[22] But I'm easing that way, my tentative intentions strengthened in part by this Tweet from the critic and playwright Terry Teachout: *Nothing is more incapacitating than the desire to be respectable (unless it's the desire to be liked).* And then, on a day when I was ready to retreat entirely, this Tweet from writer Heather Havrilesky: *Pro Tip: If you want to write something entertaining and memorable, you have to stop being such a mincing little equivocating chickenshit.*

And not just about sex. About shame in general. Its power, and its paralysis.

Alrighty, then. Away we go.

*   *   *

---

[22] He did have his limits, both confessed and not. In an introduction to his essays he said he only went as far as "respect for the public has allowed." He also admitted his honesty had practical limits: "When serving princes it is not enough to keep a secret: you need to be a liar as well." Finally, his forthrightness has been called into question by a number of writers. Sarah Bakewell quotes Rousseau referring to him as a "dissembler" who "portrays himself with defects, but gives himself only lovable ones."

In writing about sex Montaigne does a lovely job of calibrating all expectations without wrecking the fun. If he casts activities involving "the middle region" as "the sole true pleasure of human life," he also casts them simply as part of the recipe for a *full* life.[23] In a sentence where he begins by allowing that sex can be "a vain pastime," "indecorous," "shaming," and "wrong," he concludes by describing it as "health-bringing," "appropriate for loosening up a sluggish mind and body," and that it should be prescribed by doctors to "postpone the onset of old age."

By the time Montaigne wrote these lines, he was fifty-three and felt his sex life was on the decline, a fact he met with a mixture of regret, reality, and humor. "Even those to whom old age denies the practice of their desire, still tremble, neigh, and twitter for love," he wrote, but, "If [a woman] can only do [me] a good turn out of pity, then I would dearly prefer not to live at all than live on charity." But this very waning was what allowed him to write coolly on a hot subject.

When he was a young lover Montaigne overindulged in sex. His exact words were, "I burned myself at it in my youth." This is more than a metaphor, as he managed to

---

[23] Writing in the *New Statesman*, Jonathan Bate said Montaigne counted sex, friendship, and reading as the three best things, but chose reading as the best of those.

pick up a couple cases of the clap during the course of all these—as the Florio translation so delightfully puts it—"licentious allurements." He also used sex to assuage his grief after the untimely death of his dearest friend:[24] "Having need of a powerful diversion to disengage me, by art and study I became amorous."

Ultimately Montaigne's early promiscuity led him to a form of moderation. Referring to his early excesses, he says, "this whiplash has since been a lesson to me," and it shows, for in all of his writing about *l'action génitale*,[25] Montaigne rarely comes off as a bedpost-notching blowhard. (In fact, he enjoys laughing at those who are, including the newlywed who boasted he had "ridden 20 stages" on his wedding night—a lie so egregious it was grounds for annulment of the marriage.) While he admits to telling a few tales in his youth:

In my day the pleasure of telling of an affair (a pleasure scarcely less delightful than having one) was conceded only to such as had one single faithful friend.

---

[24] Etienne de La Boétie. A judge, writer, and philosopher, he died of the plague at age thirty-two. Although I make only fleeting reference to him in this book, his influence on Montaigne cannot be overstated and their relationship comprises vast swathes of Montaigne scholarship.

[25] Gosh, sometimes you don't even need to *know* French.

. . . he bemoans the deterioration of discretion in his present age:

> . . . *nowadays the most usual talk at table and when men get together turns to boasting about favours received and the secret bounties of the ladies, who really do show abject baseness of mind to allow such tender gifts to be thus cruelly hunted, grabbed and plundered by men so ungrateful, so indiscreet and so inconstant.*

The nearest he comes to a boast is in a description of his own father as "being both by art and nature cut out and finished for the service of ladies." *Ba-da-bing!* But for the most part he is unsparing on himself, whether ruefully reporting his anatomical shortcomings ("Nature has indeed treated me unlawfully and unjustly"),[26] his performance ("I have the failing of being too sudden"), or the effects of age on his penis ("It is certain that mine may now properly be called shameful and wretched").

Some of Montaigne's most entertaining sex writing is on the topic of impotence. He distinguishes between the physiological and psychological variations, with most of his recommendations focused on the latter. In one case, he

---

[26] Starobinski says Montaigne had "a humiliating phallic deficiency."

provided an underperforming newlywed pal a medallion attached to a ribbon and advised him to knot the ribbon around his midsection and repeat a mystery word three times—Montaigne didn't for a minute believe the ribbon would help, but rather that his friend's trust would instill the necessary hydraulic confidence. Indeed, it did the trick, but you can hear Montaigne snickering between the lines.

Montaigne also described psychological impotence as contagious, telling the story of a fellow who, having up 'til then been virile enough, was so put off his game by reports of another man stricken by softness that he developed the same issue:

> . . . and from that time forward, the scurvy remembrance of his disaster running in his mind and tyrannising over him, he was subject to relapse into the same misfortune.

The remedy, according to Montaigne? Lowered expectations! The man took to declaring his affliction beforehand "to the party with whom he was to have to do." Having thus taken the pressure off:

> . . . he was at leisure to cause the part to be handled and communicated to the knowledge of the other party, he was totally freed from that vexatious infirmity.

Montaigne's understanding of the psychological component of erectile dysfunction puts the lie to his earlier claim that we are no different than barnyard animals. He knew there was something more than physicality in play. He understood the aphrodisiac powers of emotion and imagination.

By his own report, Montaigne's "shameful and wretched" member still showed up for work even as he "staggered to his 50th year," but he found it a comical miniature soldier:

> I hate to see it, for one inch of wretched vigor that heats it up three times a week, bustle about and swagger with the same fierceness as if it had some great and proper day's work in its belly: a real flash in the pan.

Elsewhere Montaigne recommends that any man suffering "flash-in-the-pan" syndrome delay "this hasty and headlong pleasure" by exercising "languor in dispensation," which I assume is translated French for *Whoa there, Sparky.* And if this fails "to halt its flight . . . even in the critical moment," Montaigne says, then, "think about something else." I wish he had been more specific, as this was back before baseball.

Jousting, perhaps, although the imagery is metaphorically complicated.

*   *   *

Perhaps the most surprising element of Montaigne's essaying on sex was how he endorsed women's right to have some, even demand some. Montaigne was clearly no feminist.[27] He describes women's "essence" as "pickled in suspicion, vanity and curiosity," and the irony and jaundice with which he writes and deploys quotations in "On Some Verses of Virgil" presage the worst of this our mansplaining age. So do presumptive lines like this, written after Montaigne joked about men's impotence but then said of women: "The time is always right for them, so that they will always be ready when our time comes along."

In a chapter of *Montaigne's Discovery of Man*[28] titled "The Young Hedonist," Donald Frame says:

> *Like so many of the ancients and his own contemporaries, he generally regards [women] as potentially decorative lightweights, incapable either of good sense or of mental or spiritual elevation.*

---

[27] Beyond the scope of this chapter, you may be interested to read more about the complex relationship between Montaigne and "protofeminist" Marie de Gournay, whom Sarah Bakewell says was "by far the most important woman in his life."

[28] Gendered irony of title noted.

And yet shortly afterward, Frame writes:

[Montaigne] concludes that men have been unfair to them and kept them from their rightful equality.

In the introduction to his translation, Screech makes reference to Montaigne's "statements of anti-feminism," and deploys that blighted qualifier "in the context of his times" (at best a phrase of amelioration rather than liberation) but it is worth noting Screech's summation:

The conclusion of Montaigne is an arresting one: women should be allowed more freedom: men and women share a common "mould"—both have the common form of human kind. And that is nowhere more obvious than in our sexuality.

Montaigne observed that while women are subject to sexual passions every bit as urgent as those experienced by men,[29] they are expected to resist them while men are allowed to go right at it "without blame or reproach":

---

[29] In her blog series on reading Montaigne, Stefanie Hollmichel suggests there may have been a double standard within this double standard: *There is a bit of the "women as sex fiends" element here and throughout the essay that patriarchal society so loves using to turn women into animals or slaves of [S]atan.*

Our women can see, can they not, that there is no merchant, no barrister, no soldier who does not drop what he is doing so as to hurry and get on with "the job"—no porter or cobbler either, however weary with toil or faint with hunger.

Sarah Bakewell says that unlike most men of his time, Montaigne understood that women knew plenty about sex, and that "their imagination leads them to expect better than they get." In light of Montaigne's teensy tally-whacker, that's an especially poignant quote. The blame, he felt, lay less with the ladies than graffiti artists:

What mischief is done by those enormous [genitals] that boys spread about the passages and staircases of palaces! From these, women acquire a cruel contempt for our natural capacity.

Whatever disadvantages he might have been operating from in the sack, Montaigne held himself and men in general to account. If he commented about a disengaged partner ("Sometimes [ladies] go to it with only one buttock."), he followed up by saying "she may have a better stomach to your muleteer," which translated means her disinterest may be the product of regular and better service from the stable boy. And if, midway through said desultory session, the lady shifts to greater vigor, Montaigne suggests the man reserve his self-congratulations and rather ask himself "What if she eats your bread with the sauce of a more agreeable imagination?"

Which is to say, memories of the muleteer.

"There is no nobility in a man who can receive pleasure where he gives none," wrote Montaigne, putting all us gents on notice. The conscientious lover will study the craft; equip himself with technique and knowledge:

> *The more steps and degrees there are, the more height and honor there is in the topmost seat. We should take pleasure in being led there, as is done in magnificent palaces, by divers porticoes and passages, long and pleasant galleries, and many windings. This arrangement would redound to our advantage; we would stay there longer and love there longer. Without hope and without desire we no longer go at any worthwhile gait.*

It was in this spirit of sexual self-improvement that I clicked on the Twitter hashtag #YouAintHittinItRight. Turns out the stream was short on actionable advice, but I did it for nobility.

*   *   *

Let us depart from sex and turn to farts. I still have trouble saying the word aloud. It was forbidden in the home of my childhood, although we were free to use the phrase "passing gas" for medical purposes. There was also zero joke-farting allowed in the house. My father never once said, "Pull my finger," and even my brothers and I never ripped one in one another's presence. In fact, the only close relative I ever heard audibly flatulate was my Grandpa Perry, a gentleman in all other regards, and he only did it on fishing trips. He would always look at us perplexedly from the back of the boat. "Did you hear a buck snort?"

My circumspection in sexual matters was a full-on burlesque compared to my reticence regarding bodily functions. This is odd in that as a farm boy I spent my formative years standing—sometimes barefooted—in manure, and as a nurse I dealt with every imaginable secretion and excretion in straightforward fashion.

But lightly? Socially? It just wasn't done, and the hangover continues.

I remember the delirious giggles when Uncle Shotsy took time during the Peterson family reunion to teach me and my grade school cousins the unofficial jingle for Carter's Little Liver Pills: *Mrs. Carter's Farter Starters!* Uncle

Shotsy delivered this poetry in the voice of the Queen of England, which made it twice as hilarious.

He then launched into a round of "Beans, Beans, the Magical Fruit."

Oh, how our sides ached.

Montaigne would have given Uncle Shotsy a run for his money. He happily reported about a character "who could make his behind produce farts whenever he would."[30] (I knew that guy in junior high. I'd share his name, but what if one of the students in the school where he is now the principal read this?) Montaigne also invoked a man "that could break wind in tune with verses recited to him," which makes me wonder if the actor Jim Carrey had been reading the *Essais* when he hatched the butt-talking scene in *Ace Ventura: Pet Detective*. Cite your sources, sir.

Finally, Montaigne reported on a fellow afflicted by a "stormy and churlish" rear end who farted "constantly and unremittingly" for forty years. Based on my experience at volunteer fire department meetings, this man has many descendants.

Over a decade of marriage now, and I have not once intentionally broken wind before my wife, but if pressed I

---

[30] This is from the Screech translation. Others have it as a bowel movement; I've found none that invoke the term "shart."

will admit this is more out of embarrassment than cour-
tesy. How silly is that, considering what married folk can't
help but know of each other? "Kings and philosophers shit:
and so do ladies," wrote Montaigne, trying to ease me into
reality (Anneliese has made the same point), but I'll tell
you what finally loosened me up: two beautiful daughters.
From diaper changing through potty training, the idea
that we are ineffable spirits kinda gets sullied up. Here are
these cherubs, emitting Beelzebub butt noises. And then
you begin spending so much time with them that etiquette
is overridden by comfort. Specifically, you are reading
Goodnight, Moon for the thirty-seventh time and the kid
just will not go to sleep, and it's like you have the Hinden-
burg pressing against your spleen, and as the cramps hit
you recall Montaigne's line, "How often a man's belly, by
the denial of one single puff, brings him to the very door
of an exceeding painful death," and so, for the sake of your
health and in honor of the passage in the *Essais* in which
Montaigne praises the Emperor Claudius, who "gave liberty to let fly in all places," you give in.

"*Daaaaad!*"

But the kid's face is lit with delight, and it's teehees all
around. The comedy factor is indisputable. Nothing's fun-
nier than Dad tooting. And I recently discovered I ain't
the only one in our family who has evolved on this front.
My brother Jed, a rugged man not given to fancy behavior,

has nonetheless always stuck to the no-farting rule in my presence. And so it was a fascinating revelation the day I was babysitting one of his preschool daughters, and forgetting I was in the company of company, let the buck snort. Without skipping a beat, my niece looked up at me, chortled, and said, *"Daaaaad!"*

I called my brother immediately and said, Buddy I know what you've been up to.

*   *   *

Sex and toots, it's all good silly fun. But I am straining Montaigne for more than giggles and titillation. I'm attempting to refine my understanding of shame. Its power and its paralysis. And not all of my shame is centered on what Montaigne called the "genitories." While serving on my hometown fire department I had just backed my tanker up to the pumper when the house fire we were battling exploded in a terrifying flurry of booms, bangs, zips, and zings. Everybody on the lines spun away from the structure and sprinted for cover. The flames had eaten into an ammo cache (standard inventory around these parts). Projectiles buzzed through the air. Bob, one of the retreating firefighters, broke stride and pitched face-forward into the grass. The pumper operator was holding the siren wide open in the universal we-are-not-kidding signal for *everybody out*, but I remember thinking, *Bob!*, and then

I turned my head and tipped my helmet toward all the *bang-bang* and ran straight to him. I pulled and tugged at his shoulders, sure he was shot and bleeding. Relief swept through me when he struggled to his feet and we lumbered to safety behind the pumper. Turns out he hadn't been shot—he had simply tripped—but it was nice to look at myself in the mirror the next morning knowing I had, in that snap of an instant, run toward trouble to help a friend.

Then came the afternoon I was third man in on an interior attack and the lead man fell headfirst into a basement full of flame. I could hear Matt, number two on the hose, scream, "RIC IS IN THE BASEMENT!" I was consumed with horror. I was on all fours, but there was so much smoke I couldn't see the floor. I remember thinking *two of us will never get him out*, and then, guiding myself by the feel of the hose went scuttling backward like a panicky crab until I saw a faint rectangle of light. I jumped up and staggered out the door, smack into the chief. "RIC IS IN THE BASEMENT!" I hollered. "WE NEED MORE HELP! AND A LADDER!" Then I wheeled and dove back into the building and crashed into Matt, who was running out of the house with Ric right behind him.

Turns out, when Ric fell, Matt had hold of his legs, and was able to wrestle him back to safety. Ric said the flames were rolling all around his face mask. As we caught our breath back by the pumper, I felt relief.

I also felt shame. Later, when the three of us talked it over, I tried to explain: Believing the stairs had collapsed and Ric was deep in the basement, I had made the split-second decision to bail for more equipment and help. Matt and Ric agreed, but I wondered if I detected reservation in their voices. I know I detected it in mine.

Shame in these instances is not necessarily face-reddening. It manifests in a smaller, more slender icicle of internalized doubt: *Which way will I run?* In his essay "On Fear," Montaigne provides numerous examples of men undone by the emotion, but also leaves room for the opposite:

> . . . *even among soldiers, where fear ought to be able to find very little room, how many times have I seen it change a flock of sheep into a squadron of knights in armour . . . Sometimes fear . . . puts wings on our heels; at others it hobbles us and nails our feet to the ground.*

Subsequent to Ric falling in the basement, we were called to search the local school after a bomb threat. I geared up and went in with the rest of the crew, feeling that swagger you get when you're at the center of the action, but then someone slammed a locker door and I jumped three feet. Afterward a bystander said we were brave to go in there. I smiled, because I know that sometimes I *am* brave . . . but I have also come to understand exactly what Montaigne meant when he wrote, "the action is commendable, not the man." In one instance we charge into danger, in another we flee. Previous performance does not guarantee future results, and "one gallant action, therefore, ought not to conclude a man valiant."

Shame. It can drive us to burrow into darkness, or claw for light.

* * *

In a recurring book tour nightmare I am found dead in a Super 8 motel surrounded not by cocaine dust and scorched heroin spoons but doughnut powder and empty candy bar wrappers. Of all my secretive shames, gluttony is the most cringeworthy. Sex, well, at least most people understand that drive. And you have regenerative human emotions at stake. But to find yourself beside some freeway in Illinois watching infomercials at 2 a.m. having eaten an entire pound of M&M's *and* the whole bag of cheeseburger-flavored potato chips again, that's some low-rent hedonism. It's also a betrayal of Anneliese, who strives daily with beans and crisp vegetables to save me from myself.

Montaigne provides both solace and guilt on this front. On the one hand, he admitted to remarkable food binges:

'Tis indecent, besides the hurt it does to one's health, and even to the pleasure of eating, to eat greedily as I *do; I often bite my tongue, and sometimes my fingers, in my haste.*

. . . but he is more easily dissuaded than I:

They whose concern it is to have a care of me, may very easily hinder me from eating anything they think will do me harm; for in such matters I never covet nor miss anything I do not see; but withal, if it once comes in my sight, 'tis in vain to persuade me to forbear . . .

I'm even weaker than that, because I'd have to replace "sight" with "mind." Once something's in my head, I can't let it go. I can be on the freeway making good time, but if I envision those glazed pecans at the Love's truck stop along I-94, I'll be stopping, and they'll be all gone before I get home. Even if Anneliese purges the house of processed sugar bombs, eventually I find the chocolate chips hidden behind the frozen peas, mix them in a bowl with walnuts and brown sugar from the baking drawer, add a pat of butter, lightly microwave the works, and *voila!* Spoonable candy bar! Come the bright light of morning these recurring incidents hang around my neck not as dietary failings but character failings.

Even my one self-control outlier has evolved into a testament to my own weakness. I am a lifelong teetotaler. Never even a drop of beer. This began in service to God, then evolved into a stubborn desire for self-control. In short, when I watched as my contemporaries got drunk, it wasn't so much that I was against it, I just didn't *get* it. I maintained this position from my teens through my mid-thirties, when, wallowing in the trough of a sustained depression, I was trudging upstairs to force myself to work and a small voice said, *This is why people drink*. I paused on the landing. For the first time in my life, I was ready to try it. Just to catch a break. Just to dissolve the dread.

I stepped into my writing room. It was ankle-deep in discarded candy wrappers, vacant bakery boxes, and empty plastic cookie trays, seeded throughout with crusty coffee cups. It occurred to me I hadn't left the house in daylight for over a week. I imagined all the carbo-sugar trash replaced with empty bottles. *Nope*, said the small voice, and that was it.

Twenty years on and I still haven't had a drop, a remarkable record given my utter lack of self-control elsewhere. Emerging research proposing a genetic link between sugar cravings and alcoholism has only firmed my resolve.

Montaigne did not decry alcohol, but he did decry "gross and brutal" drunkenness, especially its deleterious effects on sex. I have spared myself that one shame.

*   *   *

There is one word that is more difficult for me to utter than *fart*, and that word is *masturbation*, about which

Montaigne once said, "That which he does unwitnessed, he does often." No comment, and why would I.

But that reticence is damaging. I was well into my adulthood when I had a long conversation with a working-class older man whose character had long ago established him as one of my few heroes. In the course of our conversation he revealed that his decades-long marriage was in fact an endurance of misery. "I married her because I wanted to have sex," he said. While this may seem ludicrous in the context of our times, we had been raised in the same church, and within that context his statement made utter sense. Then he looked at me with a mix of shame and shyness. "I wish I'd have known about the *M-word*."

It took me a minute. Then I blushed too. Because I did know about it. My Lord, how much trouble is alleviated by a few moments of time alone. And how dishonest we can be about it. I used to work out of an office overlooking a porn shop. I'd alternate between snickers and scorn when I watched guys hunkered down behind cars, waiting for the traffic to pause before darting across the street and through the plywood-paneled entryway. And yet should I happen to find a "dirty" magazine along the road in springtime (when the snow melts to reveal a season's worth of litter), I would huiltily (my personal neologism: *happily* meets *guiltily*) spend some time with it before discarding it by some camouflaged means. Then of course came the

Internet, and all of a sudden at game night you're playing Cards Against Humanity and your eminently respectable Republican West Point-graduated friend lets slip that he knows the definition of *bukkake*[31] and you want to fall off your chair except that you know what it means too, and instead silently remind yourself that neither of you learned about it at Bible study.

Oh, I knew about the *M-word*. But I had to figure it out myself. It was the one area of sex talk that seemed to leave my mother uncomfortable, and whenever it came up in school it was in terms of derision, like *who would ever do THAT?* I wish someone had been as frank about it with me as Montaigne, who recommended it as a fine means of dispersing sexual passion. In his essay "On Diversion," he quotes Lucretius in the frankest of terms—*Eject the gathered sperm in anything at all*—and then adds, "I have often tried it with profit." He is less straightforward about women's activities, but he does quote Martial saying women often achieve satisfaction "without testes to testify."

Am I embarrassed to write about this? You bet. Are you embarrassed to read it? You may already have dropped this book and headed off to wash your hands before requesting a refund.

* * *

---

[31] If you don't know, don't google.

Time came when I abandoned the abstinence track, and I shall not put specifics on it other than to say having spent most of the first three decades of my life upholding the concept of carnal acts as sacrament, it might have been better for all involved if I'd have read Montaigne on the subject a touch earlier, specifically the passages in which he cheerfully describes sex as a "ridiculous," "absurd," and "mad" activity, in which "our delights and our waste-matters are lodged higgledy-piggledy together," and at its conclusion "has something of the groanings and distraction of pain."

Instead my earliest forays were bound up in remnants of faith and weighted with unattainable ceremony. It is mortifying to recall. Desire wrapped in guilt makes a furtive bundle. There were failures and embarrassments, false starts and farce, and needless emotional contortions. All precipitated by me, a man of otherwise broad life experience and yet in the words of Montaigne, acting like "some gawky gentle dazzled youth, still quaking before his wand and blushing at it."

It was what it was. Over time it got better. Then became quite one of my favorite things. But at first shame ruled the day, shame ruined the fun.

What a shame.

By and large my sexual reticence is in remission. But there are delayed effects. I recently wrote a play about firefighters.

In one act, a character reads from *The Acquisition and Control of Fire*, by Sigmund Freud. It's typical Freudian stuff, overwrought pee-and-peener talk every which-way. In pre-production it occurred to me that as long as we had fire-fighting gear on hand, we might get a laugh if we backed the reading with a strategically rear-projected silhouette of a helmeted firefighter slowly raising a hose. We tried it out in rehearsal and fell all over ourselves laughing. But I waffled, knowing there would be people in my audience who wouldn't approve. Who wouldn't expect such crassness from me. Then I OK'd it for the premiere. Then went back to worrying about it. Yammered about it to Anneliese. Lost sleep. Sent out on social media some pre-apologies referencing "ribaldry," a word last seen around 1878.

Then came opening night, and the scene in question.

The audience *roared*.

Over fifty years old, and at all that laughter I felt something break free. As in, *of course*. Why not? The so-called "dirty joke" acknowledges our shared human experience. Our shared *silly* human experience. Cutting ourselves free from the idea of sex as sacrament.

This does not mean I've bellied up to the endless boffing buffet. "Dearness gives relish to the meat," said Montaigne, who lobbied for temperance in all things, sex included. "I look upon it as an equal injustice to loath natural pleasures as to be too much in love with them." Just because I moved beyond shame doesn't mean I'm ready to swing nonstop. Or be ribald without regard. Lately I have read several pieces examining the unnatural nature of monogamy. I am open to the idea that it is unnaturally confining. I also know that in my present, there is something to the exclusivity and trust that is essential to the point of overriding any hedonistic advantage in dispensing with all restraint. Also, marriage is about a host of other things.

In our recent popular history, whenever I heard people tittering about erstwhile professional football quarterback Tim Tebow's famously virginal state, I couldn't join in, because I recognized my young, sincere self.[32] I'm glad I'm not bound by the same sex rules as I used to be, but I also know that once you try it, unless you are utterly inured, you learn that in part you were right: It is not, as writer and critic Jessa Crispin once put it so succinctly, *a benign act*. And just because I no longer rank it as sacrament doesn't mean I wish to squander it. As Montaigne once approvingly noted of Socrates, "temperance with him is the moderatrix, not the adversary of pleasure." In overcoming shame, I do not wish to become shameless.

* * *

[32] I couldn't throw a football either.

I am told that one of Montaigne's favorite words—*bestise*—is impossible to translate. It means silliness, but also stupidity and animality. The sort of word you might associate with fart jokes. Sex. Or scarfing a whole bag of Funyuns. A word leading to shame of high or low order. And therefore an essential human part of us.

Shame, said Montaigne, "has a kind of weight." It precipitates introspection. Stripped-down, stripped-away introspection. Montaigne spoke of having both shame and respect for the person in the mirror, that it was key to a "private truth" forged in confrontation with ourselves. "The worst of my actions and qualities do not appear to me so evil as I find it evil and base not to dare to own them," wrote Montaigne. "If a man does all for honour and glory what does he think he gains by appearing before the world in a mask, concealing his true being from the people's knowledge?"

In our bathroom, the toilet faces a mirror, providing me regular opportunities to contemplate my true mortal self in a true mortal state. I can do no one else any good if I pretend some perfection I do not possess. Every base thing I've ever done has led to my being humbled by the unavoidable truth of myself. It was while "crawling on the slime of the earth," Montaigne wrote, that he noticed "the loftiest of certain heroic souls." From our humbling grows compassion, the good fruit of shame.

MARRIAGE

PROXIMITY, WITH ME, LESSENS NOT DEFECTS, BUT
RATHER AGGRAVATES THEM.

While those first two commas lend Montaigne some interpretive wiggle room, I'm just going to go ahead and read the line without the punctuation and with my wife in mind.

If I were her, I wouldn't put up with me.

I'm not saying that in the ain't-I-a-scamp, shoot-you-the-elbow, knowing-nod sort of way. I'm saying that in the *No, seriously,* sense.

Here is a story I have told previously (the dialogue has morphed with time but the punch line remains the same): A woman who had recently heard me speak at a public event encountered Anneliese in the store. "It must be *wonderful* to be married to Mike," the woman gushed.

"Ermmm . . ." said Anneliese, clearly too love-struck to respond more specifically.

"He's so *funny*!"

"Yes he is," said Anneliese, who, possessed of her own comic timing, paused a half-beat, then, as if finishing the woman's sentence, added, ". . . on*stage*."

I cannot contest the distinction.

Life with me can be miserable. As a husband I am loyal, faithful, and true in the fundamental sense, but fundamentalism is rarely about fun. Around the house, I am a moody grump and barking dad, prone to sulking and obsessive anxiety, impatient with everyone including myself (although far more indulgent of myself). I can burn three days crafting a loving description of a single spirea blossom but pass those same three days stepping over but failing to observe the laundry basket in the bedroom doorway.

Framed in an essay, this is fodder for chuckly anecdotes.

Framed in the bedroom doorway, it is just fodder.

Drawing definitive matrimonial conclusions about Michel de Montaigne is a sketchy business. It's not that he didn't say much about marriage—he did—but in regard to his own, we get only hints. We know he married in his early thirties, doing so, Donald Frame says, "on his father's urging and without enthusiasm, as a social obligation." Citing his "unruly humors" and the fact that he hated any sort of bond or obligation, Montaigne wrote:

> *By my own design I would have fled from marrying Wisdom herself if she would have had me. But no matter what we may say, the customs and practices of life in society sweep us along. Most of my doings are governed by example not choice. Nevertheless I did not, strictly speaking, invite myself to the feast: I was led there, brought to it by external considerations.*

So the honeymoon must have been a real humdinger. Still, as a man of duty, Montaigne was resolved to endure the marriage without complaint, as evidenced by this note of connubial surrender: "It is no longer time to kick when we have let ourselves be hobbled." Just get in the minivan and drive, pal.

For all his self-revelation throughout the essays, he provides very few specifics about—in fact, he rarely mentions—his wife, and never refers to her by name. Françoise de La Chassaigne was a decade his junior and the daughter of a colleague in the parliament. We get only the most granular insights on how he felt about her in the day-to-day. In a passage from an essay in which he was examining how we can hold conflicting emotions in the same heart, Montaigne reveals that he alternately looked "bleakly" and "lovingly" at his wife. In another passage, he describes how he ordered a horse be provided for his wife when he saw her struggling and stumbling on a steep road. This kindness is tempered by his admitting he was insensate at the time (he had just been in his near-fatal riding accident) and does not recall giving the order.

When Montaigne was complimentary of his wife, it was usually in terms of "entrusting" the management of his household to her. After noting the type of wife of which he did *not* approve:

> *I see, and am vexed to see, in several families I know, Monsieur about noon come home all jaded and ruffled about his affairs, when Madame is still dressing her hair and tricking up herself, forsooth, in her closet: this is for queens to do, and that's a question, too: 'tis ridiculous and unjust that the laziness of our wives should be maintained with our sweat and labour.*

. . . he then hints that his wife gives him no such trouble:

> *No man, so far as in me lie, shall have a clearer, a more quiet and free fruition of his estate than I.*

Over the course of time Françoise bore him six daughters—only one of whom survived childhood. After the death of one of them at age two, Montaigne—who was away in Paris at the time—came as near to poignancy with his pen as I have seen. Sending Françoise a translation of a letter Plutarch wrote to his wife upon the death of their child, Montaigne prefaces it with a dedication hinting that he is willing to defy the custom of the day in respect to showing her affection:

> *MY WIFE, You understand well that it is not proper for a man of the world, according to the rules of this our time, to continue to court and caress you; for they say that a sensible person may take a wife indeed, but that to espouse her is to act like a fool. Let them talk; I adhere for my part the custom of the good old days . . . Let us live, my wife, you and I, in the old French method.*

He then tells her he is sending her the Plutarch because, "I have none, I believe, more particularly intimate than you." Surely this speaks something to esteem (although the "I believe" clause is an interesting qualifier, but let's give Montaigne the benefit of the doubt and assume he was just working the rhythm of the line). The tenderest words come when, in regard to the content of Plutarch's letter, Montaigne tells his wife that he is "regretting much that fortune has made it so suitable a present [for] you." Certainly there is gentleness here. And proof that Montaigne was not gallivanting with his family out of mind. Then again, it is one thing to be home alone mourning a deceased child; it is quite another to drop a card from Paris. And in fact, in the end Montaigne the great writer commends his wife not to his own words for solace, but rather turns her over to Plutarch:

But I leave to Plutarch the duty of comforting you, acquainting you with your duty herein, begging you to put your faith in him for my sake; for he will reveal to you my own ideas, and will express the matter far better than I should myself.

Elsewhere in his writing Montaigne says "we do not require much learning in our wives," and then cites the example of a duke who, when told that the princess he was considering to marry had been brought up simply and never taught to read, was pleased, as "a wife is learned enough when she can tell the difference between her husband's undershirt and his doublet." In another *wink-wink, nudge-nudge* moment, he writes, "That man knew what he was talking about, it seems to me, who said that a good marriage needs a blind wife and a deaf husband."

My intuition here is that Montaigne included bits like this more for his pals at the pub than for himself. No less an indictment, I suppose, but I'm willing to assume there was a slight nuance. Ultimately, Screech seems to best position Montaigne's relationship to his wife:

> *Marriage he conceives of course as Christians did: as a mutually loving union of two unequals, each with duties to the other, each helping the other until death them do part. For either to neglect its duties, for either to regret or neglect its rightful pleasures or those of its partner, is to fall into the sin of ingratitude.*

Dutiful, then.

* * *

I've gone through a lot of wedding rings. Right now I've got two of them. The one I wear most often belonged to Anneliese's grandfather or great-uncle. We're not sure. It was in a drawer in the farmhouse when we moved in. My other ring—my backup ring—came from a head shop and cost nine dollars. I bought it to replace my original wedding ring, which was my stepfather-in-law's wedding ring from his first marriage.[33] I lost that one while delivering a breech lamb, and there are pretty high odds it actually came off inside the sheep. As an aside I can tell you that

---

[33] Some might see this as fiddling with kismet; in this case frugality overcame superstition.

if you loudly declare that you lost your first wedding ring inside a sheep, you can get everyone in the bar to stop their beer halfway to their face.

So I don't care terribly much what sort of ring I wear, which is good, considering how fast I was going through them there for a while. In fact, as I tell you this story now, I'm reminded that I lost yet another wedding ring in a fire truck. I removed it on my way to battle a blaze and forgot to put it back on. Perhaps it still rattles around the defroster en route to chimney fires.

I was a happy—or at least a relatively contented— bachelor for a long time. But I can tell you that if I am not especially attached to the ring as an object, I don't take the significance of the ring for granted. I don't remove it lightly, and I don't put it back on without a moment's pause to consider all it signifies.

Right now our marriage is at a critical point. We are in one of those stretches where familiarity and function are both a boon and a beware. We have become good at running our show, at keeping things together. But we are not tripping the light fantastic. And in another ten years? When the children are gone, and we are no longer bound by the mission?

We worry about this, and rightly so, per Montaigne: "Marriage requires foundations which are solid and durable; and we must keep on the alert."

Before we got married, Anneliese and I met with a minister and took a test called the Premarital Personal And Relationship Evaluation (PREPARE!). It consisted of 165 multiple choice questions, and we scored as "Very Dynamic." I think of that sometimes when we are dragging saggy-eyed past each other in the bathroom. When we walked out of the PREPARE! test I felt we were walking shoulder-to-shoulder in every sense. I knew there were no guarantees, but I knew we had a solid start. As I characterized it at the time, *four good tires and a clear windshield.*

Four good tires. Not five minutes ago Anneliese texted me from the service garage where she was getting quotes on new tires for our twelve-year-old fambulance.[34] I have been holed up in my room over the garage most of the past month, hammering away at deadlines. *Very dynamic.*

"The meaning of marriage begins in the giving of words," wrote Wendell Berry in his essay "Poetry and Marriage: The Use of Old Forms," and for the first four or five years of our marriage we sat down on our anniversary (or within a few days of it) and reviewed our wedding vows. It wasn't always fun. We weren't always hitting the mark.

---

[34] I always swore I would never own or drive a minivan, and so I don't. It's a fambulance. A man has his pride. I'm also slowly letting go.

But it was also this reminder that we are in this together. That we gave our words.

Three years have passed since we last pulled out those vows. I've spent more than one anniversary solo in a Super 8.

We're rowing the boat, but sitting apart.

*   *   *

In some of his writing Montaigne allows for the idea of love and tenderness in marriage, which sounds nice on the face of it, and one hopes he picked some flowers in that regard, but if you extrapolate from "love and tenderness" to "fulfilling sex life with the wife," you'd do well to temper your expectations. According to Starobinski, Montaigne viewed "marriage" and "amorous relations" as separate categories, and if not separate only cautiously overlapping. Consider the following ringing endorsement:

> *A good marriage (if there be such a thing) rejects the company and conditions of Cupid: it strives to reproduce those of loving-friendship. It is a pleasant fellowship for life, full of constancy, trust and an infinity of solid useful services and mutual duties.*

"Solid useful services" is a pretty good summation of Montaigne's precepts for sex between a husband and a wife:

Marriage is a bond both religious and devout: that is why the pleasure we derive from it must be serious, restrained and intermingled with some gravity; its sensuousness should be somewhat wise and dutiful.[35]

Mainly Montaigne believed that the point of doing it with your spouse is to make a baby, and that a man shouldn't have coitus with his wife if she is pregnant or too old to conceive. He reported that whenever he got it on with Françoise he "always went the plain way to work" because "debauched tricks and postures" and "voluptuous," "unremitting" pleasure can be "deleterious to the sperm," and may cause the wife to "exceed the bounds of reason."[36] Now I can tell Anneliese I'm not underperforming, I'm guarding her sanity. Montaigne also said that "a wife ought not to be so greedily enamoured of her husband's foreparts, that she cannot endure to see him turn his back." Somehow I doubt he ever shared that one with Mrs. Montaigne. I know it has led to some hoots in our house.

---

[35] In fairness, anyone married with children understands that "wise and dutiful" could fall under the umbrella of "Quick, before *Martha Speaks!* is over!"

[36] If a husband wanted to get kinky, Montaigne approved, as long as it occurred extramaritally. He defends this double standard by saying that in having a mistress he *honors* his wife by taking his debauchery elsewhere.

Out of respect to my wife and the fact checkers, I am not going to speak of our sex life with any specificity other than to say it's the usual Venn diagram in which the overlap waxes and wanes and the circles aren't always the same size. Montaigne addressed this age-old vexation when he cited the case of a man whose wife complained of her husband's "too frequent addresses to her." The man was requesting a minimum of ten "courses" a day. Seeking a ruling, the couple beseeched the queen. Her Highness decreed that six times a day was a "legitimate and necessary stint."

This may tell us more about the queen than the husband.

Whether things at our house are proceeding at a "legitimate and necessary stint" depends on who you ask and when, but I am not as glum as Montaigne, who believed "matrimonial duties" were "feeble by nature . . . if such there still be." In the essay "On Some Verses of Virgil," Montaigne quotes lines from the poet that are thrice as hot as anything you'll find on Pornhub but then casts them as "a little too passionate" for married folk, where "within that wise contract our sexual desires are not so madcap; they are darkened and have lost their edge."

But let us depart cynicism and visit hope: My pal Al used to be a leading customer at a local bar run by a husband and wife who had been tending tavern in tandem for decades. The pair was the very personification of their profession: faces seamed and cured by a lifetime of late hours and cigarette smoke, skin pale from lives lived in low light, and a permanent hunch from leaning over the bar to catch orders. The wife's bowling ball body rode on chicken-skinny legs often clad in polyester shorts, and she spoke with a nicotine rasp. The old man was of similar build and voice.

Came a day when Al had cause to borrow something from the couple's home. The husband—creaking along on his bad hip and stiff pegs—admitted Al and then led him down the hallway. As they passed the open bedroom door, the old man paused, hiked the drawers sagging off his concave butt, sniffed manfully, and nodded toward the king-size bed.

"E'yup. The ol' workbench!"

*   *   *

The first time I laid eyes on Anneliese she had just rounded the corner past Big Jim's Sports Bar in Fall Creek, Wisconsin. She was holding the hand of a bundled toddler and—although I didn't know it at the time—was headed for the local library where I was scheduled to speak. As she and the little girl passed, I noticed they had the same pale blue eyes. Now the little girl is sixteen and six feet tall and her mother and I are in our twelfth year of marriage. I never saw this coming.

Anneliese is more educated than I. She is bilingual. My second language is dropping my gerunds. She is ten years younger than I, but she is the grown-up in our relationship.

I trot that last one out a lot. Do it as a bit during monologues. Say she keeps a manila folder labeled REALITY, and every now and then has me peek in there.

Everyone always chuckles, and that's what I'm looking for—chuckles.

But it's true. When I read a passage by Sarah Bakewell in which she says Françoise "was more alert to practical concerns than Montaigne," I imagine she and Anneliese sharing manila folders and plotting to cut the dead weight.

In the Cotton/Hazlitt translation Montaigne is quoted saying, "I require in married women the economical virtue above all other virtues." The Screech translation chooses "sound housekeeping" rather than "economical virtue," so I may be flouting definition, but the first time I read the Cotton/Hazlitt I smiled.

I married that woman.

Setting aside for a moment her immeasurable contributions as executive director of our family and business operations, Anneliese supplements our family income by working as a yoga instructor, fitness instructor, and Spanish translator. She further supplements our income through her frugality. Nothing exemplifies this more than how we approached our latest auto purchase, the aforementioned "fambulance." Our previous van, which we'd bought well used, was twelve years old with a high-digit odometer and failing transmission, so it was time for a replacement. Anneliese took the lead, calling dealers, setting up test drives, and working online listings. Wadding up the weekly shopper to light the morning fire, I'd find it annotated with underlines and circles and lists of pros and cons. I mostly just provided security when we met Craigslisters in the Wal-Mart parking lot.

In the end, she found a low-mileage beauty at a low-end price. It was also the same age as the one with the failing transmission. And so if there is ever to be a moment that distills why I love my wife, it was when she traded in our twelve-year-old van for another twelve-year-old van. Reverse conspicuous consumption, and it's *still* racking up the miles. The only drawback was that it's of a model and bland color of which they made a million. You end up trying two or three doors in the Farm & Fleet parking lot before you get the right vehicle. Then the driver's side door handle broke. The salvage yard guy said the only one he had was black. I don't care about that, said Anneliese, and the price was right. "In marriage, alliances and money rightly weigh at least as much as attractiveness and beauty," wrote Montaigne, and every time I walk through all the bland vans directly to the one with the black handle, I agree and refresh my affection.

It is Anneliese who sees this family as a whole. Keeps the pantries full of home-canned. Makes sure we sit down together for supper, shut everything off, and talk. She schedules the family meetings. She is the formalizer, whereas I navigate via feeling and intuition and procrastination. When I overdo it with the children, go reactionary and drill sergeant and make rash declarations, it is she who mitigates. Waits until a quieter time and teaches me to reconsider.

This is not to paint Anneliese as some translucent spirit of ineffable perfection. But if you think I'm going to list her faults and failings A) pull the other one, B) they'd fit on a T-shirt tag, and C) I would only suffer by comparison. I am simply privileged to be at her side in any circumstance. Including, oddly or not, signing our last will and testament. We sat at the table with the attorney and made plans for our demise, be it together or separately. And when we walked out into the parking lot holding hands I felt freshly close to her, as if we had done something very grown-up. It felt ceremonial.

And perhaps the oddest comfort, but the deepest comfort? Anneliese knows my darkest secrets and innermost shames. There are no surprises for her. In my darkest hours this gives me peace of mind, lends me safety. With her I have nothing to hide.

There were two towers in Montaigne's castle. His office and bedroom were located in one. His wife resided in the other. It is not the same here, but there are parallels. Right now I can look out the window of my room over the garage and see the house where on countless evenings the family turns in while I burn the late-night lights to hit a deadline. Or fritter away a deadline. I never tire of holing up out here. If I could I would do it every day of the week, all year long. Although so as not to get in a rut, I would take an enforced break every November, leaving the writing desk to spend nine days on the back forty in my deer stand. Where I would do some writing.

What a blessing, what a fortune (even in the absence of fortune) to not only discover what we most love to do, and even more, be *free* and *allowed* to do what we most love. But what about the collateral effects upon those who must live with what we love? In *Montaigne in Motion*, Starobinski tells us Montaigne so loved to retreat to his room and write that he declared he might well have preferred to produce a child "by intercourse with the Muses [rather] than by intercourse with [my] wife."

After five deceased children and one preface to Plutarch

mailed in from Paris, you have to think Françoise might have been fine with that.

*   *   *

In addition to writing, I also tell stories for hire, do one-man shows in regional theaters, and sometimes sing, but never dance (this is a failure of neurology at the cellular level, as even the beloved polka escapes me, and yes, I know it's just one, two, three). I also sell T-shirts. (We discontinued the coffee mugs as too many were breaking in transit.) All these extraneous activities create what your big-shot corporate types refer to as "diversification," and barring tenure or a blockbuster, it is the key to survival. The upshot of this itinerant cobble of self-employment is that I am on the road somewhere around eighty to one hundred days per year.

I fancy the phrase "on the road," as it conjures tumbleweed horizons, wistful taillights, and muraled tour buses dieseling through the night. In fact, it's generally just me in the family van with a box of books and a late check-in at the Janesville Super 8 after totally rocking the public library. Still, away is away, and during these stretches I develop teen-level absence crushes on Anneliese. Become outright sappy. Call her to talk from some coffee shop when she's knee-deep in laundry and health insurance statements, teaching a yoga class, or in the bleachers at a volleyball game.

Although you couldn't blame her, she doesn't block my calls. However, we have developed the Inverse Phone Time Rule: the longer the absence, the shorter the call. Otherwise there is this tendency to slide into a form of subtle one-upmanship in which either party winds up ticking off accomplishments as a means of gently assuring the other that neither is living out of a picnic basket. Much better, we have found, to say Hello, I love you, Good-bye, and save the business meetings for the kitchen table.

Montaigne did a lot of traveling after his retirement, and when his friends criticized him for it, he countered by saying the best time for a man to leave his house was after he had put it on a course of continuing without him, by which he meant, "I put my wife to't, as a concern of her own, leaving her, by my absence, the whole government of my affairs." In the Screech translation, he says his leaving "enables" his wife to learn governance. A selfless favor on his part so that she might sharpen her "sound house-keeping" skills, because "the most useful and honourable knowledge and employment for the mother of a family is the science of good housewivery."

What he didn't say—and what I have experienced—is that among the lessons a traveling man learns is not only did the house function fine without him, it functions *better* without him. I have learned to calibrate my reentry.

Generally by the time I'm within an hour or so of home I have managed to re-imagine myself as the Conquering Hero Returned (yes, that was me, cruising northbound in a Toyota van full of empty book boxes and Ozzy Osbourne's "Mama I'm Comin' Home" rattling the defroster vents) and am therefore perplexed when I fail to find the family waiting at the mailbox, confetti cannons charged, ready to huzzah my homecoming. Rather more often there is the sense one is a rock plopping in the peaceful pond. After a few pinched feelings, a new pattern has emerged: when I am an hour or so out I call, simply as a means of getting back on the radar and factored into the calculus of the day. When I do arrive, it is into the house for a quick kiss and howdy, then off to the office for an hour or two before truly coming home. Happiness hinges on decompression. During a recent interview with a women's magazine, a traveling writer was asked if there was anything special he did for his wife when he was away from home, and I replied, yes, I don't call her and tell her how to run things.

Out on my tours, when I speak of Anneliese from the stage, or sing a song I've written about her, I get thick-throated and misty. I wish I could say I retain that emotion all the way back home, but even if it is not so dramatic when I step back over the front doorsill, I recognize myself in Montaigne when he writes, "These interruptions fill me with fresh affection towards my family, and render my house more pleasant to me." And in the translation by Screech, there is this, lovely as any verse:

> *Loving affection, as I know, has arms long enough to stretch from one end of the world to the other and meet—especially conjugal love, for it comports a continuous exchange of duties which reawaken our memory of the tie.*

Sometimes up there on the stage, I touch the ring on my finger. The way I used to every twenty minutes when we were first married.

Just to reawaken my memory of the tie.

* * *

Montaigne generally rejects the idea of married love and leans more toward friendship. But we know from his comments elsewhere he found even that friendship limited. As historian and author Dr. Paul Rahe ventured in a recent radio interview:

> *Montaigne doesn't think it's very likely true friendship can be found in marriage. He [Montaigne] says apart from being a bargain where only the entrance is free, its duration being fettered and constrained depending on*

things outside our world, it's a bargain struck for other purposes. Within it, you soon have to unsnarl hundreds of extraneous tangled ends which are enough to break the thread of a living passion and to trouble its course.

Then, in a humanizing moment, Dr. Rahe added, "Thank God that's not my marriage!"

Is it mine? I believe not. But when couples say their spouse is their best friend, I wonder what that's like. Anneliese is my truest confidant, my one lover, my utterly trusted partner. But friendship? I struggle with that one.

Montaigne did classify his wife as one of his *amis*—those he loved inclusive of friends, parents, children—but this is still short of calling her his best friend. His best friend was Etienne de La Boétie—in fact it was he who translated and left to Montaigne the copy of Plutarch that Montaigne sent to his wife from Paris with the note claiming he had none more intimate than her. Screech says it was unusual for a man to recognize his wife in this fashion in those times, so this seems evidence that Montaigne did hold his wife in some esteem.

And yet: intimacy and esteem are not friendship. And even were they, Screech also writes that Montaigne found no example of a woman with whom his sexual attraction and his friendship attained equal strength. Where both heat and warmth could intersect.

*   *   *

I am having a mid-life crisis, though not of the hair-plugs-and-a-Corvette sort. More of a look in the mirror and ask, What have you done? Calling myself into account. And nowhere more than in the case of my wife. In a perverse way, Montaigne's dismissive portrayal of his wife—and wives in general—is good for me. As much as I revel in recognition when Montaigne writes something I vibe with, when it comes to shared shortcomings, well, that's not so much fun. Especially where my marriage is involved. I want to look away, move on, turn the page. But I must not. Whatever I recognize of myself in him requires severe attention and redress. I intuit what I can about the state of his marriage and think, *I better focus*. Recently I read the Jim Harrison poem "Vows" and it knocked me on my butt, as it should have. Paraphrasing, will the evolution of my marriage be the evolution of my greatest failure? When it comes to the maintenance of matrimony the probationary period is perpetual. Even when it's sunny you know darkness will come. There is always that moment of doubt. That moment when you're not sure you'd do it again.

That moment when you see the other person thinking the same thing.

"I have in truth more strictly observed the rules of marriage than I either promised or hoped," wrote Montaigne, defining lukewarm self-congratulations. Better in this case that I look to Wendell Berry, and his idea that we speak our vows into a future that is unknown: "But that it is unknown requires us to be generous toward it, and requires our generosity to be full and unconditional."

Generosity. How I struggle with that one.

Generosity. A form of reverence. *Reverence*, a word I insisted be in our vows. A vow I violate every time I let slip other words harsh or hasty. Or when I retreat, saying nothing at all.

There is work to do. Again I turn to Berry:

These halts and difficulties do not ask for immediate remedy; we fail them by making emergencies of them. They ask, rather, for patience, forbearance, inspiration—the gifts and graces of time, circumstance, and faith.

Three days ago the children of the local elementary school—including our younger daughter—held an art show in the local library, in the very room where my wife and I first met. I got there early. We were in another hectic stretch. Coming and going. Running the show, but not tripping the light fantastic. Arriving in separate vehicles.

But then Anneliese walked in, just jeans and a light shirt, a scarf about her neck in a shade that electrified her eyes, and I saw her just as I had all those years ago.

A gift, a grace.

* * *

An ironic intermission: Beyond some fact checking, Anneliese does not read my books. She does, however, often lie in our marriage bed right beside me, reading Wendell Berry.

* * *

Not to put you off your lunch but if you snag your wedding ring just right it will strip the skin clean off your finger. Emergency medical manuals refer to this as a "ring avulsion." Farmers, mechanics, and manual laborers are most vulnerable, although I have witnessed it in more domestic settings. My friend Eric Teanecker lost his matrimonial digit when he jumped from a speedboat and caught his ring on a tarp snap. I wince even thinking about it.

So it is that whenever I clean the chicken coop, or split firewood, or sharpen the chain saw, I remove my wedding band and hang it on a pin stuck in the bulletin board beside the telephone. There was a time in my superstitious past when I would have been overwhelmed with all the implied bad karma in this action, but these days I just want to retain my finger.

There is also a contemplative opportunity: when I look down at my hand and see the groove running the circumference of my finger, I like the idea that we are working on something here that goes beyond jewelry. In summer, when the indentation is rendered even more starkly by a strip of untanned skin, I like the idea that removing the ring fails to remove evidence of the ring.

But lately it also occurs to me to wonder: when my wife removes her ring, and considers her empty finger, what does she think of the impression left?

AMATEUR AESTHETICS

**IF A MAN COULD DINE OFF THE STEAM OF A ROAST,
WOULDN'T THAT BE A FINE SAVING?**

Perhaps you failed to assume, but I owe the sublimated bulk of my aesthetic construct to Prince Rogers Nelson, circa *Purple Rain*. The film and album were released the summer after my freshman year in college. I sat solo through the movie a minimum of four times, wore the hubs off the soundtrack cassette, draped my bedroom with purple scarves, stocked the dresser top with fat candles, and scotch-taped fishnet to the drywall above the bed (intended to create seductive shadows of mystery, it wound up a pointless cobweb). I furthermore spent time scissoring words and letters out of magazines and taping them around the edges of the bureau mirror to re-create Prince's lyrics in the style of a hostage note, phonetic shorthand included (Prince was text message before text message).

Prior to this, my idea of interior design was a pair of antlers and a linked chain of all my used football mouth guards dating back to seventh grade.[37] But then came Prince. And my perception of masculinity, of beauty, of my own Midwest, expanded. Expanded a tad more than was sustainable, as it turned out, but expanded. Prince was not my single motivator, but he lit the incense. Within a year of watching *Purple Rain* I bought my first army surplus trench coat, rode a Greyhound bus out of Wisconsin to work as a cowboy in Wyoming, made my first trip to Europe, and began experimenting with hair mousse.

All us cosmopolitans gotta start somewhere.

I am here to meander (via Montaigne) through the concept of aesthetics but want to begin the trip with an examination of the aforementioned term "cosmopolitan," which is often misinterpreted as being synonymous with frequent-flyer miles and elitist smuggery. This misinterpretation is fed and fattened by doofuses who crack jokes about cowboys, Europe, and hair mousse. Having recently google-stumbled upon an article in *Encounters on*

---

[37] I am not lying.

Education, I am happy to expand my own understanding by quoting Columbia University Professor David T. Hansen, who defines cosmopolitanism as "a state in which a person comes to grips with and holds his or her identity (or identities) in a kind of generative or productive tension with those of other people." Such as when my buddy Ric texts me on opening morning of deer hunting season to accuse me of drinking green ginger tea in my tree stand and is right on.

Montaigne's work vibrates with these tensions, whether he is describing his childhood as a nobleman's son raised by peasants, or the gastronomic differences between nations: "We avoid wine from the bottom of the barrel; in Portugal they adore its savour: it is the drink of princes." In another of my favorite Montaigne essays, "Of Cannibals," rather than denigrate the practices of "savage" indigenous peoples he suggests his fellow citizens with their faith-driven wars and abandonment of the poor are far more culpable for the horrors they propagate, inasmuch as they do so while claiming to be civilized: "I am not so shocked by savages who roast and eat the bodies of their dead as by those who torture and persecute the living." Zinging his ethnocentric neighbors one last time, he points to the "savages" and says, "Ah! But they wear no breeches . . ."

This being a silly and lighthearted chapter, we will speak

no more of cannibalism. And we will—for now—set aside cosmopolitanism. We shall, however, fiddle around with generative tension right through to the end.

*   *   *

Michel de Montaigne's writing is permeated with aesthetic minutiae: He preferred a fresh napkin with each course of dinner and small, clear drinking glasses he could toss off in a single gulp. He preferred to write letters on plain paper rather than fancy stationery. He preferred clothing sewn so the seams were not visible. He wrote of the subjective nature of beauty via the Platonic sphere as compared to the Epicurean pyramid. He happily cast judgment on the aesthetics of others, writing worriedly of "delicate and affected" men who "caused their hair to be pinched off" and complaining about the return of a hairstyle favored by the ancient Gauls in which they "wore their hair long before and the hinder part of the head shaved," which I take to mean that Gauls invented the reverse mullet and Montaigne may have trend-spotted the original emo band.

In another observation with contemporary relevance, Montaigne wrote:

There are places where they wear rings not only through their noses, lips, cheeks, and on their toes, but also weighty gimbals of gold thrust through their paps and buttocks.

Yes. We call it the mall.

In another province the youths bored holes through [the penis] in public, prised gaps between the flesh and the skin and then threaded through them the longest thickest skewers which they could stand.

I had to wait for the Internet to be invented before I learned about that one.

Noting that everyone was wearing their doublets "as high as their breast" one season, then "slipped down betwixt their thighs" the next while "laugh[ing] at the former fashion as uneasy and intolerable," Montaigne found it difficult to keep up:

. . . our change of fashions is so prompt and sudden, that the inventions of all the tailors in the world cannot furnish out new whim-whams enow to feed our vanity withal . . . there will often be a necessity that the despised forms must again come in vogue . . .

I myself have lived long enough to witness the resurgence of bell bottoms, clunky eyewear, mullets, and overwrought mustaches. Just lately my teenaged daughter tells me high-waisted jeans are making a comeback after a decade of hip-squishing butt décolletage. I shall have to rummage around my keepsake drawer and gift her with my jelly bracelets, pink hair dye, and silk magenta headscarf.

Let's Go Crazy, Prince said, and for a while there, I did.

*   *   *

For over two decades I kept a clutch of my grandfather's fountain pens in a box, vowing to one day replace the bladders and use the pens to compose fungible verse. Then, in a fit of practicality and having shifted them from place to place for nearly two decades, I sold the whole batch via eBay to a man in Australia. Shortly thereafter a writer friend gave me a fountain pen as a gift and I so enjoyed the sensation of writing with it that now my desk and backpack are stocked with several different models. They require constant capping and re-capping, they dry out, they smear, they leak whenever I fly . . . in fact, one of my fingers is blotched right now. The console of my car contains more random freebie ballpoints than I can use in three years. But how lovely it is to compose with ink that flows rather than rolls.

My favorite coffee mug was given to me by a reader. It has a rubberized base, a stainless steel barrel, a ceramic rim and interior, and tapers slightly from butt to lip. The rubberized base is practically intended to keep the mug from sliding across the dashboard of a car, but I like it because it deadens the tamp when I place the mug on a hard surface. The stainless steel is visually pleasing in terms of its precise machining, implication of durability, and diffuse reflections. The ceramic lends heft and best presents the coffee to the nose and tongue. The cylindrical taper creates an impression of modernism, but also reminds me of my grandma Peterson's old coffee cups. Each of these elements combine to make my coffee taste better.

It feels inane bordering on profane to discuss these things in the face of this dust-blown barb-wired world. What shallow spendthrift pays one copper penny—or wastes breath in praise—for anything beyond utility? And yet, I would like to believe that aesthetics are a form of grace, and that grace cultivates grace.

Nothing encapsulates and conveys the mysterious power of aesthetics more succinctly than the fact that I prefer it spelled that way rather than the shorter *esthetics*. There is something about the meshed *ae* that drops my blood pressure five points. I can't explain it, but I can feel it.

I'm self-conscious about that, and my ambivalence about aesthetics (or beauty for beauty's sake) is easily tracked to the blue-collar themes against which I rebound again and again, that whole business of worth being chained to tangible results in the form of piles of salable goods (hay bales) and economic expedience (I buy my socks by the bag at a big-box store). I imagine there is also some hangover from my roughneck past, where *aesthete* rhymed with *effete*, and that was all you needed to know.

On book tours I wear logging boots to remind myself that my brother Jed works daily at a vocation that has twice delivered him to the ICU—once with a fractured skull, once after a tree branch shish-kebob'd his neck. When I sit there so heavily shod while signing books in a library anteroom, I intend to remind myself of him and all those who give their bodies to labor, but those steel-toes are also a form of armor for my ego. A signal that I'm not just some art-head softie. That I can carry more than one aesthetic. Once out on tour late at night I called Jed from the road. He was somewhere in northern Wisconsin, running a load of pulp. I was southbound through Illinois, my van crammed with book boxes. "We're both just out here haulin' trees," I said. It seems I'm wired to apologize for aesthetics and legitimize art in terms of pragmatic economics.

* * *

Montaigne claimed to disdain fancy clothes and said in a perfect world we would all run around naked. Humans are animals, he reasoned, and all other animals are delivered to the world wearing everything they need. By wearing clothes, Montaigne maintained, we lost our ability to live without them. It's a lovely theory, but as a guy from Wisconsin, I have to say that if a man survives January without wearing any clothes, come February he will not require any of Montaigne's advice about sex.

The main problem with clothes, Montaigne seems to feel—and I tend to agree—is their complication beyond utility. The minute someone hacked armholes in their mastodon hide, form went to war with fashion.

It seems the only answer was to quit caring:

I have ever been ready to imitate the negligent garb, which is yet observable amongst the young men of our time, to wear my cloak on one shoulder, my cap on one side, a stocking in disorder, which seems to express a kind of haughty disdain of these exotic ornaments.

Switch out the cloak for a fire department hoodie, put a bill on the cap, and we're pretty much on the same page. Recently I was heading for town on an errand when my teenaged daughter stopped, turned, gave me and my clothes a slow once-over, shook her head, sighed, and returned to her phone. (There was no eye-roll. I accepted this as mercy.) In this her time of peak fashion-consciousness, she has been sentenced to pass through life beside an oldster dad who looks like he was taught to dress by dirt farmers and heavy metal roadies given five minutes and three bucks to shop in a low-end thrift store. Which is fine by me, and Montaigne: "I desire therein to be viewed as I appear in mine own genuine, simple, and ordinary manner, without study and artifice."[38]

It's a free country, as we are reminded by radio talk show hosts when it serves their purpose, and so it is we may arbitrate our own tastes. Low-lying fruit can yield fine wine: *Wall Street Journal* theater critic and playwright Terry Teachout reports he discovered Rossini via a Bugs Bunny cartoon, and I like to think of aesthetic fluidity as a form of egalitarianism—however, in culture as in traffic, intersection creates its own danger zones. As a junk food junkie raised in the 1970s and '80s who has nonetheless come to appreciate the odd dollop of truffled asparagus foam, I am susceptible to believing anything. One Sunday morning we were running late for a church service and I was in the kitchen power-chowing my breakfast while my teenaged daughter and her cousin did their makeup in the bathroom. Suddenly they came giggling around the corner.

---

[38] Sounds good, until you google around and find the portraits in which he is clad in silks and satins, puff shoulders, and a lace neck ruff the width of a bird bath.

"Uncle Mike!" said my niece, proffering a slim cylindrical container tipped with a plastic nozzle, "Try this!"

Understand: I had just finished my eggs. I was operating in the context of food. I extended my hand, palm up.

My niece squirted a delicious-looking dot of sparkling confection on the pad of my index finger.

"It fizzes!" she said.

Finally! I thought to myself. *The magic of Pop Rocks meets the technology of aerosol cheese!*

I popped the dollop into my mouth. Both girls gasped.

In an instant I was draped over the sink, pawing at my mouth like the dog who chomped the porcupine.

"I didn't say *eat* it!" screeched my niece. The two girls were now falling over each other with laughter.

When I finally stopped spitting and sputtering, I picked up the container and read the label: WILD APPLE DAFFODIL SHIMMER FIZZ BODY MOUSSE.

Setting aside for a moment the concussive effects of realizing that a thing called body mousse even existed (and that my daughter and I would shortly be discussing exactly what circumstances might require its application), what you had here was proof that aesthetically speaking, it is not always best to pair an open mind with an open mouth.

* * *

In the matter of aesthetics, M.A. Screech has described Montaigne as a "gentlemanly amateur" with "a natural bent for appreciating style at the expense of matter" who was happy to pursue the "delightful" over the "useful." He ranked his favorite poets for the beauty of their writing as much or more than for the power of their content. But while he advocated for aesthetics he did not consider himself an arbiter. Nor did he trust anyone who did, or made it an undue priority. Long before Kardashians roamed the land, Montaigne fretted "I hate our people, who can worse endure an ill-contrived robe than an ill-contrived mind."

Montaigne's reservations in this respect extended to the government. Even as he grumbled over his fellow citizens' cultural priorities, he defended their right to choose. He despised "sumptuary laws," citing silly regulations curtailing the wearing of jewelry or certain clothes and believed the more the government said no, the more the market for forbidden fruit grew: "The way by which our laws attempt to regulate idle and vain expenses in meat and clothes, seems to be quite contrary to the end designed." I take this to mean contemporary Montaigne would happily lobby for your right to guzzle a Big Gulp and spark up a resinous fatty. At the very least, he would have been pro medical marijuana, as he is on the record as an advocate of alternative therapies, and was in fact an aromatherapist before his time:

Physicians might, I believe, extract greater utility from odours than they do, for I have often observed that they cause an alteration in me and work upon my spirits according to their several virtues; which makes me approve of what is said, that the use of incense and perfumes in churches, so ancient and so universally received in all nations and religions, was intended to cheer us, and to rouse and purify the senses, the better to fit us for contemplation.

This makes me feel better about my affection for vanilla candles, cedarwood incense, and my essential oils diffuser, all of which help me produce essays on deadline. For purposes of my own philistine feng shui, I store the diffuser on a shelf five feet from my deer rifle. I feel this juxtaposition is justified by Montaigne's admission that high and low culture are essential to the aesthetic teeter-totter: "None of our tastes are pure and unalloyed . . . wisdom has its excesses, and has no less need of moderation than folly." I am free to savor six kale chips and an episode of *It's Okay to Be Smart* then dive into a tub of Costco Cheese Balls and a four-hour *Reno 911* binge-watch. "I now, and I anon, are two several persons," Montaigne once said, the upshot being, the "other people" necessary to achieve the cosmopolitan state of generative tension required for aesthetic fruition sometimes reside within the same body.

Each of us will set our own standards. I am at my writing desk when I receive a non sequitur text message from my brother Jed.

"*Also, neck tattoo saying F.U.*"

I text back: "*?*"

He answers: "*Ya at waterpark observing body art.*"

I can picture him, a white-thighed logger likely wearing jorts.

I call him then, and he does a play-by-play as folks cycle through the wave pool. Your average Wisconsin waterpark features a lot of tattooable acreage per capita, and it was a poignant report. We don't agree on everything, my brother and I, aesthetics included, but at the conclusion of our phone call we were patriots united in our belief that the American flag should not be sprouting back hair.

*　　*　　*

Once upon a time a banjo player in my employ took me to task for stating from the stage that I wasn't a very good musician. Self-deprecation is fine, he said, but you're overdoing it.

This threw some sand in my gears, as intended.

Self-deprecation (as expressed by "I could be wrong . . .") was the keystone of Montaigne's philosophy. He was forever reminding his readers that we really shouldn't listen to him. He warned us that many false opinions arise through "the over-good opinion that we have of ourselves," and Starobinski notes how Montaigne characterized his own moral life as "exemplary enough if you take its instruction in reverse." He also knew—as Hugo Friedrich has pointed out and my banjo player was implying—that false modesty could veer too cutesy-clever: "There is a certain type of subtle humility that is born of presumption."

Alain de Botton says all of this self-deprecation is Montaigne's lovely means of letting us know that our feelings of inadequacy are normal. I certainly appreciate this passage:

> *And, having undertaken to speak without distinction of everything that presents itself to my imagination and making use only of my own innate resources, if it happens—as it often does—that I by chance come across in the great writers the very topics that I have undertaken to discuss . . . recognizing how clumsy and slow-witted I am in comparison with them, I feel pity or contempt for myself.*

This is what I was getting at when I made those remarks from the stage. I was surrounded by musicians who knew their instruments inside and out; who had dedicated their lives to their craft; who could play with delicacy and nuance ("all the dusty parts," as we say, in reference to those areas of the fretboard the rest of us never visit). As opposed to me: in my thirties before I learned my first chord, I haven't learned many more and play my guitar with all the nuance of a man cutting brush. Every time we rehearse I recognize how clumsy and slow-witted I am in comparison with the band that has my back.

And writing? When I crack and read two pages of Dylan Thomas or Zora Neale Hurston, I understand I am in the Tour de France on a tricycle. But rather than triggering my own version of Montaigne's self-contempt, the words of greater writers leave me invigorated and thrilled that I am even allowed to be in there swinging. Rather than run from the keyboard, I am propelled back to it, determined—even in the mix of my magazine makeover pieces and my after-dinner yokel anecdotes and my library talks—to attempt a handful of beautiful things. To admire, then aspire.

When the banjo player and I talked it over later, he explained that when I overdo the whole boot-scuff act, I am implicitly denigrating the people who enjoy my work, and devaluing the time they took to attend and the dollar they chose to spend (he may not have used the phrase "implicitly denigrating"). There is this temptation, coming from where I do in every sense, to mock everything but meat and potatoes as pretension. To admit I love modern dance but quickly put my thumb on the other side of the scale with some offhand yay-hoo crack about deer huntin'. Lately whenever I invoke the word "artisanal" it tilts toward pejorative. I'm still struggling with how to remain a sensible and unpretentious fellow while celebrating the civilizing power of aesthetics. I'm striving for a point of equidistance between snark and sanctimony. Perhaps this was what Terry Teachout had in mind when he wrote that we must be careful not to become "terrible simplifiers." As a critic, says Teachout, part of his job is to accept and revel in complication. The drive to stand there in my unearned logging boots and rationalize art will always be there, as will be the bills, which cannot be paid with the steam of an epigraphic roast. But Teachout also wrote, "Those whose privilege it is to make art also have a simultaneous duty to magnify the beauty of the world." Montaigne preferred his water in clear drinking glasses because they "let my eyes too taste it to the full." When I read that passage I am reminded—taught—that a statesman, a soldier, a *pragmatist*, can value beauty for beauty's sake. That my knee-jerk need to justify my aesthetic choices—even in terms of pragmatism—is pointless. Examine them, yes, and question them, but mostly just get on with it. And get on with getting better. If Montaigne reveals his aesthetic preferences through a filter of self-deprecation, as an "ordinary" guy, he does so at least in part and in the hope that he might speak to an audience wider than just the self-appointed arbiters.

In an essay examining whether or not the appreciation of beautiful things can improve our character, the philosopher John Armstrong says that when we look at the statue of Apollo and are moved, the point is not to adopt the same pose or hairstyle, or to become an archer; rather it is to "seek to realize in ourselves the fusion of the drives embodied by the sculpture." When I saw Prince on the big screen, I felt some draw, some desire to be more than I was. To make some incremental move in his direction. The initial steps were ludicrous but necessary: The leaden-footed white boy lip-syncing "When Doves Cry" in the mirror (knowing full well he couldn't even polka in wafflestompers let alone pirouette on stilettos) was dancing—via cosmopolitan tension—toward Montaigne's most gracious, most definitive, aesthetic decree: the greatest thing in the world is to know how to belong to oneself.

KIDNEY STONE WISDOM

'TIS FOR MY GOOD TO HAVE THE STONE: THAT THE
STRUCTURE OF MY AGE MUST NATURALLY SUFFER SOME
DECAY...

The day before my fifty-first birthday I was confronted by the face of Tom Cruise. He was glossy on the cover of a grocery store tabloid, and I was slumpy in the checkout line. *Jeepers,* I thought to myself, *Tom's aged better than me.*[39]

In that very instant, my groceries scrolled toward the checker. She handed me a leaflet and said, "Sir, will you be taking advantage of our discount today?"

The letters were large, and therefore I could read them:

10% OFF TUESDAYS
FOR SENIORS 55 AND OVER

---

[39] Improper grammar. Save your energy and postage. This is how I think.

"We do not know where death awaits us," wrote Montaigne, "[s]o let us wait for it everywhere."

* * *

When Michel de Montaigne was thirty-five years old, his formerly hale and hearty father was laid low by a kidney stone. And then another. And another. The attacks continued for seven years, until his father, "dragging [on] to a very painful end," died "tormented with a great stone in his bladder." Montaigne—who claimed "even from my infancy" to have held a great horror of kidney stones, found his childhood fears reinforced, his dread deepened.

He birthed his own first stone right about the same time he began writing his *Essais*.[40] Then—like father, like son—came the avalanche: "My fits come so thick upon me that I am scarcely ever at ease."

At one point in the *Essais* he classifies "five or six" of these fits as "very long and sustained attacks"; in another he casually references voiding two large stones "after supper." And yet, by the time Montaigne had done his own seven-year stint with "gravel," he had come to see it as something to which he had been "generously introduced,"

---

[40] I wonder if it was a coincidence that he developed stones after adopting the sedentary writing lifestyle. Then there was this quote: "I can hold my water ten hours, and as long as any man in health."

as it gave him a chance to strengthen his character before
witnesses:

> *They see you sweat in agony, turn pale, turn red, trem-*
> *ble, vomit your very blood, suffer strange contractions*
> *and convulsions, sometimes shed great tears from your*
> *eyes, discharge thick, black, and frightful urine, or have*
> *it stopped up by some sharp rough stone that cruelly*
> *pricks and flays the neck of your penis; meanwhile keep-*
> *ing up conversation with your company with a normal*
> *countenance, jesting in the intervals with your servants.*

Jesting? The only time my mother has ever heard me
utter the word *fuck* I was thirty-seven years old and she
was driving me to the emergency room as a stone excori-
ated the inner circumference of my spasming ureter. I had
not yet read Montaigne, and so was unaware that "men of
the best quality are most frequently afflicted with it: 'tis a
noble and dignified disease."

Dignified? I barked and whinnied, I sank to my knees, I
pounded the dashboard, I apologized for saying fuck, I dry
heaved, I said *fuck* again, I apologized again.

We still hadn't left the driveway.

In her essay "Kidney Stone in My Shoe," Sonya Huber
writes:

> *I sighed with relief when Montaigne stopped quoting Seneca and turned toward his real body, even when he dished about the details of his agony with kidney stones . . . Montaigne's kidney stones are his path to humble brilliance through the vulnerability of describing illness.*

And then, writes Huber:

> *Montaigne's decaying body was also his writing teacher. As he ages and becomes ill, he becomes vulnerable and specific.*

I am at that age where the body begins to override determination and character. Recently I took a manuscript to town to do revisions in a coffee shop only to discover that I'd left my reading glasses at home. I stubbornly ordered coffee and unpacked the manuscript anyway, then, after squinting migraine-hard from a great distance to fruitless effect, grew pouty at the realization: this will not be overcome with gumption.[41] While we'll have to agree to disagree on the ennobling power of kidney stones (in my experience they *vaporize* your moral/ethical fiber), when

---

[41] I also felt resentment toward Montaigne, who bragged about being fifty-four and not needing spectacles.

we move beyond that into the vulnerability referenced by Huber, I concur. Slowly I am rounding the corner on the idea that even a wimp such as I can draw edifying lessons from the frailties of body and mind, no matter how trivial.

I am valetudinary!

I add the exclamation point because I always thought of *valetudinary* as an exclamatory, exhortative sorta word. Y'know, like *valedictorian!* and *extraordinary!* had a baby.

Then, whoops, I got to reading the dictionary and discovered it's not an honorific, it's just a fancy synonym for hypochondriac. Specifically, a person who is unduly anxious about their health.

My being valetudinous is contradictory. I am relatively relaxed about death, but worry daily (often while sleeping too little and eating too much) about my health. Recently I met with a representative of a breakfast club called the ROMEOs (Retired Old Men Eating Out) who begin each meeting with a ceremonial "Listing of Afflictons." Should I ever be invited to join, I am refining and expanding my own list beyond the kidney stone to include cholesterol numbers that would be good if they were a major league batting average; a Morton's neuroma that swells up like a grumpy meatball on the sole of my left foot every few years; a hip that went bad to the point of scheduling surgery until a year at a treadmill desk loosened it again; a series of voice and throat issues possibly caused by gastric reflux; partial blindness in my left eye; nerve impingement to my left arm and leg with resultant weakness, tingling, and foot drop; a partially detached clavicle; a clicking right thumb;[42] your standard neck and back glitches; persistent tinnitus and hyperacusis; a bad shoulder; a benignly lumpy left testicle; and, intermittently, some gout. Also last year I lost four crowns.

Sometimes I get so obsessed with what I imagine is going wrong that I end up in the clinic. Got second and third opinions on my nerve weakness, and while all three doctors confirmed it, they couldn't explain it. Convinced I had sleep apnea, I spent a night in the sleep clinic and was given a clean bill of snooze. After running a camera-equipped tube up my nose and into my stomach the physician who couldn't quite explain my voice problems did definitively state that I had the longest esophagus she'd ever seen.

It's a helluva pickup line at the gastroenterology conference.

*   *   *

Three a.m., ten degrees, thin swift clouds and chill moonlight. I am speed-walking out our potholed driveway. The

---

[42] Ever since I jammed it in someone's shoulder pads during a football game in the 1980s.

naked branches cast jackstraw shadows. Now I break into a sprint, hitting myself in the head again and again.

Then it's back to a walk. I attempt a cleansing breath but the air hisses out of me like someone knifed a tire. I break into a sprint again. More head-hitting.

What you've got here is something I call the "freakout fartlek." *Fartlek* is a snickery Swedish word I picked up from my high school track coach. Basically it means running, then walking, then running, then walking, over and over, in short bursts. The track coach said it would help me outrun everyone in the Little Lakeland Conference two-mile, and mostly that was true. Out here in the frost-white moonlight I'm trying to outrun my own humiliating hysteria.

Back in the room above the garage there are strips of paper all over the place, scraps of a book that is nearly due and tonight I've finally realized I ain't gonna make the deadline and this is not an artistic problem this is a financial problem and a responsibility problem and back there up the driveway is a house all dark and inside it three hearts beat and they went to sleep thinking that yellow-lit window out across the yard meant Dad had everything in hand when in fact he is currently halfway out the driveway smacking himself in the noggin and blowing like a spooked whitetail.

Walk, run, walk, run, smack myself atop the head rat-a-tat

like some bongo player on speedballs. Anything to dislodge the dread adrenaline.

At the end of the driveway I stop beside the mailbox and tell myself out loud: *Get it together!*

Couldn't be more cheesy, couldn't be more cliché.

Back in the room above the garage, I blow out another breath, square my shoulders, and say out loud, *OK, here we go*.

Stoic, dammit. It's bootstraps time.

And then the anxiety spawns again, a burst sac of panic blooming downward through the chest and liver like chilled bleach.

Shameful, is all I can think.

Shameful.

When I first began writing, every session was a Lipton Tea reverie. The first article I ever sold was an essay titled "Courtin' Country Style," a light little number describing the dating habits of rural teens. The first story I ever sold to a national publication was a meditation on lawn mowers, and how their pervasive suburban buzz was a signal that all was well. One of my most enduring rent-payers has been a live-audience recording of me reading a collection of small-town tales I wrote called *Never Stand Behind a Sneezing Cow*.

In those neophyte times, I hungered for the day when I could support myself as a writer. And yet, once I quit my last "real" job (i.e., the last time someone else was paying for my health insurance) it didn't take long for me to lose equilibrium. My love for—or obsession with—writing has never waned. It's still the first thing I want to do every morning. But with the deadlines—especially after I married and we had a family to support—and the necessary isolation, I found myself mirroring Montaigne's experience after he retired to write, as described by Screech:

Montaigne's project of calm study soon went wrong. He fell into an unbalanced melancholy; his spirit galloped off like a runaway horse; his mind, left fallow, produced weeds not grass.

In Montaigne's time, "melancholy" was not simply defined as moping. It was a form of ecstasy and inspiration, and "ecstasies" were seen as wisdom. But it did have a dark side, and Montaigne, a generally content and confident fellow, found himself staring at the blank page and sliding into *chagrin*, defined by Screech as "depression touched by madness." As a nobleman with a castle and all that inherited herring-and-wine money, I assume Montaigne never bolted from his castle in the wee hours in a deadline-driven panic attack triggered by the mortgage.[43] But while I may not have servants and a vineyard I do have a multitude of privileges (being free to write chief among them), and thus when I am swept away by my own *chagrin*, I feel a self-loathing exacerbated by my Scandinavian Stoic suck-it-up background. This is weakness, Spanky. Take it to the closet and don't come out until it's smothered.

Montaigne had a better approach, as encapsulated by this summary from Michael Sperber's *Psychiatric Times* piece:

> *[Montaigne's] goal was to achieve eudaimonia (equanimity in the face of adversity). The best way to accomplish this was through ataraxia (freedom from anxiety) requiring control of the emotions and remaining in the here-and-now.*

And so when the *chagrin* struck, Screech tells us Montaigne struck back—with his pen:

> *Writing the* Essays *was, at one period, a successful attempt to exorcize that demon. To shame himself, he*

---

[43] On the other hand, he was living in the middle of a civil war that more than once came to his very doorstep, so there are limits to the parallels drawn.

tells us, he decided to write down his thoughts and his
rhapsodies.

If I understand the use of the word "shame" correctly in this context, it is Montaigne's attempt to drag *chagrin* out of the closet and face it in the light. Or it may just be an admission—similar to mine above—that these emotions are embarrassing when we feel we haven't "earned" them. Or a blend of both. I do know from reading M.A. Screech's *Montaigne and Melancholy* that Montaigne disdained *tristesse*, a "delightful delicate sadness" thought by those who suffered it (and cultivated by those who did not) to be an aristocratic sign of intellectual sensitivity associated with melancholy genius. Screech says he "took care to distance himself from the affectation" and spoke of his own melancholy in "belittling terms," calling himself instead "an empty dreamer."

During one anxiety-riddled stretch I kept feeling pressure in my chest until finally one night it got so heavy it was affecting my breathing. Convinced I was having a heart attack, I snuck from bed at 4 a.m., left Anneliese a note not to worry, and drove myself to the Emergency Room. I have often lectured other people against exactly this sort of granite-headed stupidity—in many cases while I was wearing an ambulance uniform.

I wound up getting the whole works—placed under observation, IVs, blood draws to monitor my cardiac enzymes, then a treadmill stress test—all of which I passed. This good news was balanced by the humiliating memory of my dear wife and daughters standing at my bedside looking at a crazy man who had just cost the family roughly eight grand.

There is the temptation to link myself to Montaigne and describe these emotional swings as some writerly occupational hazard, but that's a self-centered shortcut. I know of farmers with crippling depression; whereas I'm worried about missing a deadline, they're worried about losing a year's crops. If I lose my farm, I have to unload three-dozen chickens and do my typing someplace else. If they lose their farm, they lose their livelihood.

My brother Jed the logger is the prototypical backwoods tough guy. Short, quiet, and lean. Unblinking direct gaze, close-mouthed, and has proved his mettle by rousting trespassers in the dark while armed with nothing but a grin.[44] This is not a guy who spends time musing about. Closest he comes to *tristesse* is when he pops a hydraulic line on his log forwarder. And yet he fights a near-constant battle with anxiety and depression. Recently he and I were up in the room over my garage, both of us bottom-dredging a

---

[44] OK, and also a double-barreled shotgun he calls Betty Lou.

trench as deep as we'd ever sunk, comparing notes on our bouts with anxiety, and for the first time—I hadn't even told Anneliese—I admitted the part about the freak-out fartlek. As I described how I ran in fits and starts down the driveway, hyperventilating and hitting myself, trying by physical means to escape the psychological, his eyes widened with recognition and in that instant I realized he was saying *me too*.

Neither of us is the type to hug it out, and we didn't. But we had a long, helpful talk. And I was reminded sometimes we don't reveal our secret stories for ourselves but for others whose secret stories are the same—and the secret wants out.

* * *

I believe I can match Montaigne mood swing for mood swing, but when it comes to kidney stones, I have a shameful admission.

Montaigne dropped a dump-truck full.

I passed *one*. Years ago.

And *still* I pee in fear.

But I'll tell you what Montaigne didn't have: *Proctalgia fugax*.

And if he did, he wasn't brave enough to write about it, despite all his protestations about utter self-revelation.

Remember a few chapters back I plucked up my courage

and perhaps ran off half my readership by invoking masturbation? Well, that was spring geraniums compared to *Proctalgia fugax*.

Which I have got.

It first struck one night in my early twenties. I'd been asleep for just a short time when I awoke to an indescribable pain that I will describe anyway as amateur proctology performed with a red-hot and poorly grounded curling iron. I mean, we are talking a bullet-sweat, bubble-eyed, liver-quivering, bust-out-the-smelling-salts situation.

Today's Wikipedia says: *During an episode, the patient feels spasm-like, sometimes excruciating, pain in the anus.*

Here, lemme edit that for ya: *During an episode, the patient feels spasm-like, ~~sometimes~~ excruciating, pain in the anus.*

It's like you can't wait to faint.

This malicious sneak of an affliction invariably waits to strike until you've sunk deep into the first forty-five minutes or so of sleep. There is also the delicate matter of trying to avoid waking your partner while clutching your buns and groaning cross-eyed in the dark. You know you've been married awhile when dawn arrives and in the fresh light of a new day, your wife's first words to you are, "So—the butt pain?"

The only cure I've found is to draw a scalding bath and

sit in it. This is tricky, say when you're pulling ambulance duty and staying at a friend's house and they wonder why you're having a midnight soak. I'd hang a hand towel on the spigot to deaden the sound of the water splashing into the tub.

Several sources report that the number one cause of this literal pain in the ass is stress. I'm not sure, but if you aren't stressed *before* you have it . . . I trust science can find the cure, and I'm determined to raise awareness, but it's been a struggle to get someone to pose for the poster.

* * *

The confrontation and contemplation of our instabilities— physical, mental, and butt-centric—are a means of easing ourselves into the idea of mortality. In *Montaigne in Motion*, Jean Starobinski makes the point that the nobleman's overarching goal in composing the *Essais* was to write his own memorial, and thus "the whole project of writing is seen in the perspective of death." Introducing the book, Montaigne wrote:

> *I have dedicated [this book] to the private convenience of my relatives and friends so that when they have lost me (as soon they must), they may recover here some features of my habits and humors, and by this means keep*

the knowledge they have had of me more complete and alive.[45]

In composing his own literary death mask, Montaigne expressed one vain regret:

For this dead and mute portrait, besides what it takes away from my natural being, does not represent me in my best state, but fallen far from my early vigor and cheerfulness, and beginning to grow withered and rancid.

Sir, may I interest you in our 10-percent-off Tuesdays?

In his book *The Swerve*, Stephen Greenblatt writes that death was one of Montaigne's two favorite subjects (sex the other, natch), and Screech maintains that "fear of dying . . . haunted his youth and young manhood." Montaigne wrote of himself, "never did a man so distrust his life, never did a man set less faith in his duration." This is all personally and historically understandable. Within one ten-year stretch Montaigne lost his best friend (a loss he also used as an excuse to have a lot of sex, so you know, pros and cons), his dad, an uncle, a brother, and two of

---

[45] Not that he felt moved to write anything to memorialize *them*.

his daughters; he furthermore lived in a time of civil wars, religious massacres, and deadly pestilence.

In his early years, Montaigne was obsessively determined to prepare himself to meet death coolly and without complaint. He figured if he studied enough philosophy he could die like he was occupying a beautiful painting: reclined palely in artfully arranged sheets, bidding aristocratic adieu to a circle of mournful admirers, his exit scene lit porcelain white. Trouble is, if you make death your primary study, you soon understand that despite your noble strivings, death may not allow you to die nobly. As a child I too used to entertain visions of myself conducting dramatically staged departures (in particular I recall acting out a bizarre scene below the bird feeder in which I perished in grand slow motion from a hockey injury—the fact that I never played or watched hockey is only one among a truckload of interlocking non sequiturs). By the time my very real-life sister died of a congenital disorder at the age of six, these romantic necro-visions had ceased. But in the main my constant sense that death is, if not near, *nearing*, is directly derived from my experiences as a firefighter and emergency medical responder. Once you've been handed a red plastic biohazard bag and assigned to pluck femur shards stuck dart-like through a sheet metal car door, you no longer labor to sustain the illusory likelihood of sweet goodbyes, of fortuitous timing, of guaranteed anything.

In his essay "That to Study Philosophy Is to Learn to Die" Montaigne wrote, "Death can surprise us in so many ways"—then rattled off a long list of folks who perished under unusual circumstances: crushed by a crowd, stubbing a toe, bumping into a door, choking on a grape seed, using an infected comb, smacked in the head by the shell of a tortoise dropped by an eagle, taking a tennis ball to the cranium,[46] and (in a list including at least one pope) tapping out "betwixt the very thighs of women."

Montaigne came to accept that death set its own schedule and script. But it took kidney stones and horses to really set him free. Kidney stones taught him that philosophy will not keep you composed when you are trying to pee red-hot welding slag. And that runaway horse smashing into him from out of nowhere was a neat metaphorical reminder that we can't always see these things coming. In the wake of the accident, his servants and riding companions described how he lay on the ground retching and vomiting blood. And yet all Montaigne recalled was a sweet peace as he accepted what he assumed was his final breath. When his head cleared, his preparations for death relaxed, shifting from obsession and rehearsal to simple contemplation. Ultimately he said he'd just as soon die planting his cabbages, untroubled

---

[46] Montaigne's brother, age twenty-three.

by death's arrival and even more untroubled by his abandoned garden.

Once when Montaigne was returning from a trip, he thought of something (he doesn't say what) he wanted done after his death. Although he was feeling hale and healthy and was only about three miles from home, he whipped out his notebook and wrote the postmortem instruction down on the spot. "I had hastened to jot it down because I had not been absolutely certain of getting back home." Just last week I was leaving in a rush for a book tour when it occurred to me that I might not make it back. Anneliese was teaching and my children were in school, so I stopped just long enough to scribble on a Post-it note, which I left atop my office desk:

I love my little family and all the joy they have brought me, each with their own bright heart.

Then, realizing that if indeed something did happen to me the note could be misconstrued, I added a parenthetical PS:

(No portent, just mean it.)

When my actual funeral takes place, there will be a collective rolling of the eyes. But I have long harbored this compulsive need to inform those around me that if I die right now it's OK, because I have been given so much. I did it to my poor banjo player during sound check at the Thrasher Opera House two days ago, and cornered my brother John about it while we were deer hunting last November. Another time I felt compelled to pause and share these preemptive sentiments with Anneliese just as I stepped out the door to catch a flight that would take me over the Rocky Mountains and into an assignment of moderate danger. I scared her half to death and received a stern lecture upon my return.

Thanks to an article I read in the *Financial Times*, I now know what Montaigne and I were doing here was engaging in *premeditatio malorum*, a Stoic practice in which we regularly remind ourselves that we and our loved ones are mortal and bound to die. Basically it boils down to contemplating potential misfortunes in advance, so when the bad thing happens you can react with all the grim calmness of Eeyore. I think of it as pragmatic pessimism, best expressed in the original Michael Perry adage, *If life gives you lemons . . . well, that's what I thought was gonna happen.*

Death is the ultimate lemon, ever tenuous on the tree. One afternoon this past winter, finding myself chilled and drowsy after lunch, I stoked the fire, pulled Great-grandma's old recliner in close to the hearth, pointed my slippers toward the stove door, and drifted off. I awoke with a start and smoldering slippers. I'm not just alliterating here—smoke was curling off my toes, and the soles were bubbling.

To expire by inhaling the smoke from one's own slippers is, I suppose, unlikely in the extreme. And yet my first thought was of Montaigne, who said death stalks our every instant, and therefore "We should be booted and spurred, and ready to go."

But *slippered*?

* * *

I am reminded, sometimes, when the anxiety dissipates or the depression lifts or the *Proctalgia fugax* retracts its talon, of how Montaigne felt when his latest kidney stone pinged into the thunder mug:

> But is there anything delightful in comparison of this sudden change, when from an excessive pain, I come, by the voiding of a stone, to recover, as by a flash of lightning, the beautiful light of health, so free and full, as it happens in our sudden and sharpest colics? Is there anything in the pain suffered, that one can counterpoise to the pleasure of so sudden an amendment? Oh, how much does health seem the more pleasant to me, after a sickness so near and so contiguous . . .

I want to be careful with this next line of thought, because the idea of cultivating pain or melancholy (*à la tristesse*) is an insult to those who have it foisted upon them (No, *foisted* is too polite—dumped like ten cubic yards of gravel). My logger brother lost both a wife and a child in separate tragic circumstances. He doesn't need me to rhapsodize about how my down days yield up a few nice lines. I don't want to come off as some sadness tourist. (So too it is with my "listing of afflictions," trotted out for comic effect, knowing that any number of readers may be struggling with pathologies for which there is no comic relief.) Rather, I am reporting from the fuzzy border between pathology and temperament.

Screech writes that Montaigne's melancholy was just enough to give him "some modest experience of 'vehement' disturbances"; I should like to think mine is the same, and that I use it for good and gratitude. Yes, my dips and dives have led to some decent prose now and then, but most importantly (I hope) they have given me just taste enough to act tenderly toward those who wait in vain for the upturn. Or those who cannot continue.

Montaigne's writings about suicide wander off into the weeds of theology and a guy named Cleombrotus and I can't claim to be clear on his conclusions. In some cases he expressed admiration for those who took their own lives out of principle and with grace, as well as those philosophers who felt it was better to die than live in misery. He quoted Pliny (the Elder), who said "stone in the bladder" was one of three diseases over which a man might acceptably kill himself. But even as Montaigne's own stones helped him mitigate his fear of death, he asked God to prevent him from doing the job himself and cast suicide as a "vicious extreme" he would choose only if pain "dissolve[d] [my] intelligence."

How will I react if I incur debilitating chronic pain or incapacitation? As a nurse I worked in rehabilitation with people who had strokes and spinal cord injuries, and recently I had a long talk with my buddy Ozzy about his life since becoming quadriplegic and I believe the answer to the question can only be found in the experience. As I mentioned before, even in the depths of depression and at the heights of anxiety, I've never felt suicidal. I've had plenty of stretches where I wouldn't have minded if the bus hit me, but I've never gone looking for the bus. I assume this is a privilege of chemistry, because it for damn sure isn't character.

After my freak-out fartlek session in the driveway, I worked around the clock for forty-eight hours and made no progress. At 3 a.m. again, I let myself in the darkened house to try for sleep. Slipped carefully into bed so as not to wake Anneliese.

Heaved a sigh. Rolled left. Then right. My dumb bald head sheeny with flop sweat. Failure sweat. The kind of sweat that breaks out cool like rubbing alcohol, then goes straight to chicken grease.

"What can I do for you, baby?"

Anneliese. Awakened by my agitation. Tender in a way I so rarely am.

I told her about the deadline. How my head was locked up, nothing happening, no way I was gonna make it.

"I'm sorry, baby," she said, rubbing my neck and shoulders. All I could think was that I deserved no such consideration.

Then, more business-like, "When does your editor get in?"

"Nine, usually. Eastern. Eight, here."

"That's when you call her, then. First thing. Tell her you can't do it."

"But we need the—"

"We can't have you like this."

The fear fled me.

I was asleep within the minute.

One day after my brother Jed and I had our anxiety talk, I called him and he said things were going better.

"Why, do you think?" I asked.

"Ah'dunno," he said, "but I'm goin' with it."

And again we laughed with recognition.

Montaigne warns us that if we slip too far into the "cold" side of melancholy, we become "dull, stupid, timorous and sad." My brother and I recognize ourselves here: immobile, self-loathing, angry at our own face, a panicky rider in a self-perpetuating loop. And yet it is this very lability—the rebound, in which we are so ecstatic to have broken free of the sadness, that seems to launch me to that place Screech describes as "a drive on the part of the soul to leap 'outside itself.'" The freak-out fartlek is all stops and starts and you just keep losing ground. But then comes the moment when the lead leaves your chest, when the chilled bleach evaporates in sunlight, and you just run like mad, go, go, go, like you'll never slow down again. Sometimes in those stretches I type page after page, until the dawn, the great glorious binge its own form of revenge on the gloom.

I got that book done. I had to make the early morning East Coast call, and it was a humbler, and we had to shuffle some dates and family funds, but I got my breath back, and made the home stretch.

The anxiety is never far away. I still have high heartbeat nights. Cold sweats in the darkness. The worst of it always hits between 4 a.m. and dawn. I give up finally, leave the bed, brain feeling like it's steel wool wrapped around a nine-volt, but it's so much better to be doing than stewing.

And I haven't done the freak-out fartlek since the night Anneliese spoke to me in the darkness.

Sometimes I get impatient in my marriage.

Because: I am a fool.

But not such a fool that I would ever forget what is owed.

* * *

When Montaigne did die, it wasn't easy. He contracted quinsy, a complication of tonsillitis in which the wall of the throat becomes engorged with pus. Suffocation in slow motion. A horrible way to go, and in fact it seems the dread kidney stones really did get him in the end, as the quinsy was secondary to an infection arising from a backed-up kidney. He was fifty-nine years old. His quest to philosophize himself into dying stoically and heroically had long ago morphed into a more moderate dedication to living a full and rich life, as critic Roy Porter describes:

> In time Montaigne softened, and taught himself to doubt, to reflect and accept . . . to know yourself you had to be yourself . . . The early "To philosophise is to learn to die" became, six years later, "it is philosophy that teaches us to live."

It turns out—based on deathbed reports—he did die honorably and stoically despite the horrid quinsy. His younger self would have been proud. But I long ago switched my greater loyalty to the late-stage Montaigne. The one who warned me never to fall into "lazy health," but rather to live knowing a kidney stone can drop me like I'm shot; knowing that for all my outward sunshine I have run helplessly through the night; even, cripes, knowing that a ridiculous cramp in the ass can render me unable to function. What I owe my minor illnesses (or even the illusion of illness, as I walk out of the cardiac lab with an embarrassingly clear bill of health) is self-examination, and an acknowledgement of my own vulnerability. And then to leverage that vulnerability to deepen the vigor of my life by—here I paraphrase Montaigne—countering its brevity by appreciating its weight; compensating for the speed of its flight with the strength of my grasp; and moving booted, spurred, or slippered toward death in the spirit of these, some of his loveliest words:

I have seen the grass, the blossom and the fruit; and I now know their withering. Happily so, since naturally so.

MEDITATING ON FAITH

**PRACTICE MAKES US SEE AN ENORMOUS DISTINCTION
BETWEEN DEVOUTNESS AND CONSCIENCE.**

I receive the news in a yoga studio in a strip mall on the outskirts of a mid-size Wisconsin town: Mingyur Rinpoche has gone missing.

It was only this morning that I learned to pronounce his name.

Across the parking lot there is a Burger King.

Life is a whopper.

*   *   *

Michel de Montaigne was christened a Catholic, took an oath of fidelity to Catholicism at the age of twenty-nine, and claimed the denomination to his grave. I toss

---

the qualifier "claimed" in there because reams of paper, centuries of colloquy,[47] and many million pixels have been expended in defense, doubt, and even dismissal of his faith. If you wish to wander through those weeds (*fideism*,[48] anyone?), I commend you to the experts. Some contend he was a covert atheist. Burke classified him as "no ordinary Catholic." Elsewhere he is described as a "very original Christian sceptic [*sic*]" who leaned toward reason, but only if it was touched by grace. In one bio he is cast as "a conservative and earnest Catholic," but with an "anti-dogmatic cast of mind." Another refers to him as "skeptical but sacramental." He may have doffed his cap to the Pope with one hand, but the other held a fistful of humanism, pragmatism, and skepticism. Screech points out that Montaigne specifically referred to his *Essais* as "unresolved fantasies" that sought, rather than declared, the truth, and that only the Church had that authority. More and more toward the end of those essays (and thus the end of his life), Screech says, Montaigne "professed complete submission to the Church of Rome," a take

---

[47] Can never spell that one right the first time, and not confident pronouncing it in public. Using it here in the British sense of "an informal conference on religious or theological matters."

[48] Reliance on faith rather than reason in pursuit of religious truth (per *Merriam-Webster*).

reinforced by Frame, who says Montaigne's loyalty was to Catholicism over all. Finally, in both word and deed, Montaigne himself made it clear he did not want to run afoul of the Church.

From up here in the cheap seats, I submit my completely amateur opinion: Montaigne was a doubting Catholic covering his bases. And yet: whatever the state of his faith, it informed his every move.

I would say the same for me.

*   *   *

I was raised in an obscure fundamentalist Christian sect. I enjoy announcing this because it makes people uncomfortable. There is the image of my hunkering within a walled compound, hoarding diesel fuel and fertilizer. (We did hoard diesel fuel and fertilizer; we used it to plant corn.)

Ours was a non-denominational bunch, with no official name beyond those assigned by outsiders or for purposes of government paperwork. Informally, some within called it "The Truth," and we described ourselves as "The Friends."[49] The sect was New Testament based, and came out of Scotland sometime around 1900. We eschewed churches and

_____

[49] Nope. Not the same group as Garrison Keillor. I get asked that one a lot.

met in each other's homes to share hymns, prayers, and homemade homilies. We did not celebrate Christmas.[50] Women were to wear modest dresses (no pants) and keep their hair unshorn and "up" in buns pinned atop their heads. Although many elements of his experience were different than mine, it resonated deeply when I read an interview of former Blackpentecostalist Ashon Crawley in which he said:

> *I grew up very committed to the church, loving every-thing about the world even in its insularity, even in its enclosed, peculiar nature. In fact, we relished being "a peculiar people."*

Yes, *peculiar.* The loveliest sort of peculiar. Nestled beneath His shelt'ring wings, as the old hymn had it.

It felt good to be so. *Safe* to be so.

Like we were in on The Secret.

And then one day (we are pressing the one-decade fast forward button here) you realize you no longer believe as you did. It dawns on you that revelation may be unreliable. Now you are spiritually wary.

*　*　*

---

[50] Due to pagan overtones. Fundamentalist Christians keeping Christ out of Christmas.

In my hometown of New Auburn, Wisconsin (population generally in the neighborhood of 500), when the funeral of a Lutheran or a Methodist or whatever ya got draws too many people for the respective pews (or the person is not "of church"), the Catholics over at St. Jude's open the doors to their larger space and let the service be held there. In Montaigne's time, Catholics and Protestants were cutting each other's guts out.

As a public official and private citizen Montaigne often found himself called to negotiate between both sides.[51] He expressed disgust with the whole works and came to distrust things like "ecstatic revelation" and spiritual rapture. I know something of that ecstatic revelation, even if it was a low-key Scandinavian variation in which I committed my life to Christ in the basement of the Moose Hall in Barron, Wisconsin, during the concluding verse of "Close Thy Heart No More." For an instant, right here on earth, I felt the pure power of the deity as a shaft of light beaming unbroken from heaven to my heart. There was the idea that I could be drawn weightlessly up it, free of these Moose Hall bonds.

Some look to the light always.

Some look away.

---

[51] "Both sides" is an oversimplification: included were all the usual internecine sectarian, political, and royal intrigues, in which just because someone seemed to be on your team didn't mean they weren't gunning for your seat. Montaigne navigated all of these.

*   *   *

I married into yoga. In fact, the first photograph I ever saw of Anneliese was one of her holding some pose for a newspaper article on the subject. I discovered the photo while googling her prior to our first date. The line between due diligence and online creeperism is slim.

Over time I've taken a stab at a pose here and there (it turns out yoga is not really designed for "taking a stab") and have even attended a few sessions with an instructor. I slip easily into quiet time on a wooden floor, but it's the regular return appointments that seem to elude me. As I have stated elsewhere, yoga requires patience, dedication, and follow-through, and I just don't think that's fair.

I have also practiced yoga at home, to similar intermittent effect. There is an impatience in me that wants to begin the morning with coffee at the keyboard; twenty minutes of reflection at sunrise drives me sideways with restlessness. There is also a self-absorbed selfishness factor in play, but let us save that for other times and essays, because in the scene as it is set, I am attending a two-day meditation retreat with my wife. It is possible to simultaneously settle your chakra and solidify your marriage. *Advised*, even.

According to the brochure I've skimmed while on break (refreshments included tea and pine nuts and exactly zero

Slim Jims or Ho-Hos), the meditation we are practicing is of the Tergar variety, and the techniques themselves are the product of the mind of the aforementioned Mister Rinpoche. Actually, "Mister Rinpoche" isn't quite right because there are a lot of Rinpoches, and I'm not sure it's a last name as such.

Earlier, as our session commenced, the guest instructor shared the details of Rinpoche's disappearance. Telling no one, and under the cover of darkness, the lama left his Indian monastery with nothing but the clothes on his back. We were not to be alarmed, said the instructor, as Rinpoche left behind a letter explaining that this was a planned "extended solitary retreat," but it was expected to extend beyond a year, and his location was unknown.

The corporeal Rinpoche being unavailable, we watch him on DVD. He is sitting in the lotus position on an overstuffed chair. He seems childlike and sweet. Sometimes he giggles. He says he can help tame my "monkey mind." I'd like that, and it would be a real favor to my family. But pretty quickly my monkey mind wanders over to wondering about just where it is the worldly and the otherworldly intersect. At what point does the lama's bladder overrule his mind? Does he ever have to leave the meditation cushion and strain for a vein-popping bowel movement? Are there cookie crumbs beneath the cushion? Does the lama not stub his toe and curse, does he not have to pay the videographer's invoice? Given two, would he be able to resist the second Mallo Cup?

And whither the lady lamas?

I cannot ask Rinpoche these things, as he has disappeared, leaving us with only his image on this DVD. At intervals, the instructor pauses the player and guides us through meditation sessions. I enjoy the stillness of it. The quiet. The fact that I am allowed to do it on a chair rather than force-pretzeled atop a foam pad. The idea that I can travel so deeply into this quiet world within my head and yet just a short walk away are chicken fries. The monkey mind, it is not so easily tamed.

Later I visit the Tergar website and read Mingyur Rinpoche's note. It's nice. Humanizing as heck. It's got your searching pilgrim themes, sure, with Rinpoche humbly and beautifully casting himself as "a tiny firefly in the midst of the sun's radiance," but it also conveys his acknowledgement of the mundane ("don't worry, I'm not having a mid-life crisis") and a wink or two:

> *To help you continue along the path, I've prepared many teachings over the past few years that will be delivered by my emanations. These emanations can appear magically almost anywhere and will teach you just what you need to deepen your practice.*

An *emanation*! Now *that's* lama-worthy! Then—and you can read the grin right there in the words—Mingyur reveals that by "emanations" he means those DVDs we watched, and that:

In some ways, my video emanations are better than the real me. You won't have to feed them or put them up in a hotel. They will wait patiently until you're ready for them. And most importantly, they won't feel bad if you get bored and turn them off!

Saint, sage, or sinner; a sense of humor goes a long way.

Regarding the details of his disappearance, Mingyur says he is essentially embarking on an open-ended walkabout, and will spend some time meditating at sacred sites and in caves. At this I smile, because it delivers me to Montaigne. When Montaigne retired at thirty-eight, his intention was to meditate. He would never have gone to a cave because A) he had a castle, and B) he said we should be cautious about retreating to the wilderness because the point is not to get away from the crowds *around* us, rather to get away from the crowd *inside* us. But of course that last line merges perfectly into Rinpoche's teachings about taming the monkey mind, and furthermore in Montaigne's own words I find little difference between "cave" and "man cave": "We should set aside a room, just for ourselves, at the back of the shop, keeping it entirely free and establishing there our true liberty, our principal solitude and asylum." Yes, although retreating to your castle room (or *any* room, for that matter) to muse and scribble is a privileged option, and one must tightrope the line separating self-discovery from self-indulgence. Montaigne stated that we mustn't bind ourselves to family or possessions or even our health if these things deprive us of our solitude; the real-life issues with that line of thinking are self-evident, and my wife would like to discuss the state of the screen door.

While I certainly cherish my room above the garage—and he clearly cherished his room in the castle—Montaigne also believed solitude was portable, that "you can enjoy it in the midst of towns and in the courts of kings." Or inside your head. It's a mess in mine, but the mess travels. Whether it is due to having been raised in a large family or just general wiring, I am able to withdraw deep into my head in nearly all circumstances. As a shy person I sometimes go into public to write, which may seem counterintuitive but in fact self-consciousness is a catalyst for withdrawal. The presence of strangers compels me to lower the louvers and focus directly in on the task at hand.

Montaigne equated solitude with liberty. Every time I crawl inside my head I set myself free. Solitude—my *place secrète*, as the lovely phrase has it—is my church and my confession booth.

* * *

After reading Mingyur Rinpoche's letter, I read his bio:

> *When he was twenty-three years old, he received full monastic ordination from Tai Situ Rinpoche . . . During this period, Mingyur Rinpoche received an important Dzogchen transmission from the great Nyoshul Khen Rinpoche, a renowned teacher from the Nyingma School of Tibetan Buddhism. For a total of one hundred days, spread over a number of years, this great meditation master transmitted the "oral lineage" of the Heart Essence of the Great Perfection.*

Groovy.

> *These teachings on the breakthrough (trekchö) and direct leap (tögal) of the Dzogchen lineage are extremely secret and may only be transmitted to one person at a time.*

Extremely secret? Suddenly I'm grumpy.

One of the things I loved about the church of my childhood was the lack of intermediation. We had preachers, but their job was to spread the gospel, not run Sunday services. The first Sunday after I professed my faith in the basement of the Moose Hall, I read a verse from the New Testament and offered my interpretation of it. *Preached*, in other words. That Sunday, and every Sunday for years to follow.

Montaigne felt the Bible "is not for everyone to study" and interpretations should be left to official churchmen. I preferred it our way—the only human between me and God was whoever wrote up that section of the King James. Could things go wrong? Well, sure. You heard some eyebrow-raisers now and then. I talked with an acquaintance recently who left our church for a more standard Baptist version. He was glad to be free of amateur hour. People in his new church, he said, "have really *studied* the Bible." I knew exactly what he meant and why it would be reassuring. But I don't automatically trust that a human wrapped in robes and a divinity degree is any closer to the truth—and I claim Montaigne for my team when he says "Take that clearest, purest and most perfect Word there can ever be: how much falsehood and error have men made it give birth to!"

I thought of that line when I read the website of a local church: "We believe that in the Bible, God says what He means and means what He says—God is that good!" And yet a pastor is retained for the services of interpretation.

God says what He means and means what He says, but is not trusted to speak for Himself. His inerrant truths require parsing. Context, and rephrasing. For instance: *But I say unto you, That ye resist not evil: but whosoever shall smite thee on thy right cheek, turn to him the other also.* Seems definitive. But time and again even this straightforward verse is vulnerable to the parsonage equivalent of relativistic postmodern deconstructionism leading to what Montaigne called "divinatory nonsense":

> *There are so many ways of taking anything, that it is hard for a clever mind not to find in almost any subject something or other which appears to serve his point, directly or indirectly.*

Thanks to all those years spent composing Sunday morning testimonies (often, I admit, while others were giving theirs), give me any random piece of junk mail and I can cobble you up a homily. In the words of Montaigne, "Every subject is fertile to me: a fly will serve the purpose." Because of this I retain an abiding skepticism for anyone who claims to be in possession of grand secrets only they can dispense, be it Mingyur Rinpoche, the coffee-shop proselyte, or the well-lettered theologian. In many cases the subtextual language conniptions deployed to explicate the truth wind up inadvertently demeaning and diminishing the power of the

Being in question. When we claim to speak for Jesus, it is also possible we are simply interrupting him.

Montaigne could admire philosophers and sages, but he did not revere them. He believed all disciplines were fallible. He seasoned everything with doubt. He stripped away the divine from men not to demean them but to make their lessons more relevant. Elsewhere on the Tergar website, Mingyur Rinpoche is referred to as "our precious guru," and this too makes me crinkle my nose. I just can't take it seriously. I can do the meditation, I can give the teachings my honest, earnest, humble attention, I can be respectful, but I just can't buy in to the dude or the movement.

I'm not being snarky here. My old flopped fundamentalist self distrusts or is made uneasy by true believers, but the people of faith who raised me were gentle and kind. I can express my disagreement without mockery. I reserve the right to find humor in all things—religion included, Mingyur Rinpoche included—but tempered by respectful consideration. So it is I can be transported by meditation practice but still entertain an adolescent curiosity about the reality of Mingyur Rinpoche having to abandon the lotus for the loo. Universal consciousness, sure, take your shot, but it is time on the toilet that equalizes us all. Montaigne writes of Antigonus, who, in response to a poet praising him as "son of the sun," replied, "He who has the emptying of my chamber pot knows to the contrary." When I wonder about Mingyur in the bathroom, or focus on how he jokes about his DVDs, it is because I trust that part of him more than I do the "extremely secret" transmissions.

Montaigne was less concerned about finding definitive conclusions in philosophy than *utility* in philosophy. Philosophy, according to Terence Cave, was filled with "provisional possibilities to be explored, not systems to be adapted and appropriated." I say this is true of everything up to and including Tae Bo. For a while I was very much into the Existentialists. My understanding of their philosophy amounted to: The only thing I can really control is my own behavior, so let's focus on that. Lately Montaigne has gotten me interested in the Pyrrhonists, who believed in undermining cultural prejudice and dogmatic certainty by gathering as much information from all sides of an issue as possible—but then claimed you couldn't be certain of anything, even your own uncertainty. That doubt itself is subject to doubt. This sort of thinking can spin into a Mobius strip of silliness, but when applied within reason is an essential source of humility, which Montaigne exemplified by rewriting, appending, and updating his *Essais* until the day he could no longer hold a pen.

Is it possible for spirituality to mean more to us the less we believe? I dunno, but the echoes certainly persist. Harold

Bloom wrote that even as Montaigne developed a skepticism for transcendence he continued to strive for transcendent self-knowledge. I love that word, *transcendence*. I've invoked it a lot over the years. But even as reading Montaigne has caused me to reconsider my use—or affection—for the term, I feel no less inspired by the concept. Likewise, I am no longer certain of the existence of the "soul" (Montaigne thought it was intertwined with the body) and yet I use the word often to describe that part of me—of all humans—that remains apparently ineffable despite all science. I use the term "soul" because I was raised on it. You don't just discard these things. They inflect all life to follow. Faith need not curdle; it can evaporate slowly, leaving behind its lacy trace. In his "peculiar people" interview, Ashon Crawley spoke of how he still loves the music of his church:

> I kept listening to black gospel music, and still do listen to that genre more than any other. And this wasn't just because I enjoyed it as a "style," which is a term used almost always to denigrate and discard the radical potentiality and edge of the music. I enjoyed it, and enjoy it still, because the music still moves me, still resonates with me, still causes me at times to cry, still compels me deep in my flesh, still vibrates . . .

Crawley also speaks of listening to the Hammond B-3, and fellow worshippers clapping their hands, yelping, screaming, and crying out in ecstatic joy. To be clear: In our church we were having NONE OF THAT. We constrained ourselves to the *a cappella* dirge delivered in prim seated position. A sturdy woman named Florence was allowed to play an upright piano at gospel meetings, but the one time she laid down a touch of left-handed barrel-house rumble, everyone's eyebrows shot up and an elder told her to cut the honky-tonk. We kept joy in our hearts, not our feet—and certainly never let it get anywhere near our asses. But I offer this characterization in fondness, because just as Crawley does his, I cherish our old hymns. I recall them in terms of purity and yes, even joy. When I hear Iris Dement sing, "God Walks the Dark Hills," I am filled with sweet vulnerable wistfulness. When I attended a gospel meeting with my former churchmates a few years ago I found it hard to listen to the callow minister with his greeting-card theology, but those hymns? I sang them fully.

I think what some people miss about us skeptic/agnostic types is that implicit in our position is a distrust of human reason. Of its pretension. It's not that we're too smart to believe in God, it's that we're not smart enough. Or at least not smart enough to nail down the details. Nor are we frolicking heedlessly along—I consider the state of my soul daily. In fact, maybe I'd get more done if I'd just leave it to some dude on Sunday. My take on agnosticism is best put by John Horgan (the quote within the quote is by Stephen Batchelor):

> *Agnosticism is often denigrated as a passive worldview, the philosophical equivalent of a shrug. But true agnosticism, Batchelor contended, consists of the active cultivation of doubt and uncertainty in the face of the mystery of existence. An agnostic stance "is not based on disinterest. It is founded on a passionate recognition that I do not know."*

And now, in those last four words, we are back at Montaigne. If, like his Pyrrhonists, we layer doubt upon doubt, we don't stop looking. Bryan Appleyard put it beautifully: *[G]od is an essential subject of study without which you will lead an imaginatively impoverished life.* I try to be more Christlike; that doesn't mean you'll see me in church. Montaigne himself valued Christ's humanity over his divinity. "A soul's worth doesn't exist in soaring heights but in steady movement," he wrote. "Its greatness is exercised in an intermediate, not a mighty state." I care less if you declare yourself a Christian than if you are Christ-*like*. How you treating my family? How you treating my neighbors? Everything else is just hatchets and meringue.

Someone said religion is the poeticization of ethics, and in that sense, Jesus Christ is still my chief poet. Kierkegaard the Existentialist was helpful but as much fun as a chilblain. Montaigne has joined the reading late, but I'm ready for him now, and the fact that he works bawdy now and then doesn't hurt. Divinity plus failed erections. You feel like you're working with some attainable standards.

I'm less interested in confirmation or repudiation than I am in exploration. Pursuing the *lineage* of thought. Jesus helps me. Mingyur helps me. Montaigne helps me. Each is one of many measuring sticks. You use what's in the toolbox. We need a goad as much as a god.

* * *

I kept tabs, and after a few years, Mingyur Rinpoche re-emerged. Out of the caves and into the people. Starobinski says Montaigne blended his contemplative humanism with his civic humanism. Both men remind me I can hang out in the room over the garage and noodle around the keyboard trying to think myself into a better person, but I must also visit a veteran, assist at the food bank, and attend the fire meeting.

In the centuries since his death, Montaigne has wavered in and out of favor with the Catholic Church. Over that time and right to the present, volumes have been written in deep examination of the state of and nature of his faith. I am working with fat crayons. I can report that after saying good-bye to the neighbors in writing (the advance state of his quinsy had rendered him unable to speak), on September 13, 1592, he summoned a priest to say mass. At some point in the celebration, Montaigne clasped his hands as if to pray, began to rise, then fell back dead.

Many take the presence of the priest to be an endorsement of Montaigne's faith and fealty to the church. Some are more skeptical. In faith or in memory of faith, we tend to cover our bets. Anneliese and I own an old Chevy Silverado pickup truck, a good runner for trips to the feed mill or for firewood or to deliver a load of meat chickens to the butcher. In the winter we use it to plow snow. The previous owner stuck a decal on the back bumper. "GOD IS AWESOME," it says. Every now and then I'll be headed to town and think, *I oughta peel that offa there.* More than once I've knelt to do it. Invariably I think, *Well, why poke the bear*, and leave it be.

WHAT TO DO

MAN MOVES ALL TOGETHER, BOTH TOWARDS HIS PER-
FECTION AND DECAY.

Over the first few years after I discovered Montaigne,
I read him mostly for what my people call "shits and
giggles." I snickered at the naughty bits, basked in
the ol' "Ah, that is so!" when he pinned human nature in
a manner concordant with my own opinions, and nodded in recognition of those tics, traits, and tendencies we
shared. I still enjoy that aspect of his work. For instance,
I have long been disgusted by the allegedly civilized status
of the handkerchief, and maintain that nothing is creepier
than carrying your snot in your pocket. So of course I
am tickled when I come across a passage in which Montaigne not only questions the practice but endorses my
favored alternative: the farmer snort, also known as the

207

snot rocket.[52] There is vindication and entertainment in this. But as fun as it is to watch the chickens gaggle over table scraps, eventually you remember you came to the coop for eggs, and over time I realized I was settling for the most superficial self-satisfactions in Montaigne's work and overlooking a more central element: the importance of responding to change with change.

In his autobiography *Life*, Rolling Stones guitarist Keith Richards tells of the time in the 1960s he witnessed a hip jazz crowd booing when Muddy Waters plugged in and went electric:

> *[The audience] wanted a frozen frame, not knowing that whatever they were listening to was only part of the process; something had gone before and it was going to move on.*

As I type these words, I am fifty-one years old. Middle-aged, late middle-aged, whatever. Not young, not ready to flop and stop. Raised to respect my elders, I have now watched many of them grow brittle of thought and bitter of mind. It seems that somewhere around my current life

---

[52] A little trick for clearing the nasal passages that I'll leave mostly to your imagination except to say it involves laying one's finger aside of one's nose in the manner of St. Nick and you wouldn't want to do it indoors.

stage, people make one of two moves: Some stiffen, dig in their heels, and attempt to block the future; others reinvigorate life by blending it with the spirit of youth. I hope I will—and I am working to—bend toward the second. I am not talking here about the embarrassment of an oldster trying to vibe with the kids. Nor am I talking about abdicating principles. I am talking about offering a hand, opening new doors, and sometimes—when new blood is best—stepping aside and standing down. "Youth is making its way forward in the world and seeking a name: we are on our way back," said Montaigne, who felt that too much was made of mere seniority and often punctured the idea that age automatically conferred wisdom or compelled deference. Rather, it "imprints more wrinkles in the mind than it does on the face; and souls are never, or very rarely seen, that, in growing old, do not smell sour and musty."

Recently I told someone I was learning a lot from Chance the Rapper. The person responded with a chuckle, but I was dead serious. I wasn't looking to co-opt the art of this twenty-three-year-old man, or dress like him, or try to fake up some hip appreciation of his oeuvre; rather, I was observing how he carried himself as a citizen, as a partner, as a father, and as a self-employed artist. I am working on this idea, striving for this balance, by which in growing old I might benefit from some youthful osmosis. And I'll not limit myself to defining youthfulness strictly in terms of age; I'm thinking also of the flexibility of thought. I am trying to avoid the "frozen frame" Keith Richards described,[53] or, to quote Montaigne quoting Cicero, "the mind stupefied by habit."

In *Montaigne's Discovery of Man*, Donald Frame casts the *Essais* as "a book that is professedly a record of change" and that "many apparent contradictions are really changes of opinion." Montaigne himself referred to these changes as "the train of my mutations." In his memoir *Great Books*, David Denby sees Montaigne as existing in a constant state of "becoming" and thus very suitable to an American way of thinking. I suspect this is true, although the older I get the less susceptible I am to the "American way" of anything, as so much depends on who is doing the talking. Or the advertising. But as I write these lines in the wake of an election predicated on the idea that we might "Make America Great Again," I am reminded of the Dudley Marchi quote used by Hannah Brooks-Motl in her essay "Michel de Montaigne, Time Traveler":

Each successive historical period produces interpreta-tions of Montaigne in its own image, modifying a text

---

[53] Life has a lovely way of serving up context: within forty-eight hours of reading that passage, I checked my Twitter feed to find that Sir Keef was caught up in a social media kerfuffle after he characterized rap as music for "tone-deaf people" and heavy metal as "a joke."

whose collage-like qualities seem easily adaptable to
such variegated reconstitutions of their identity.

Lovely word, *reconstitution.* As opposed to *renunciation.*
Or *retrenchment.* Surely I am not alone in my weariness
of all the binary rhetorical Ping-Pong. America is a col-
lage, ever-changing. I do not—again, because of having
read Montaigne, who was writing in the waning years of
the French Renaissance, when moderates were going thin
on the ground—assume change will be benign. "Nothing presses so hard upon a state as innovation," says Montaigne in the Cotton/Hazlitt translation; Screech is more
foreboding, saying "novelty" will "crush" the state. Either
reading is thought-provoking at present. Reflecting on his
time as a magistrate, Montaigne lamented that "laws are
often made by fools, more often by people who, in their
hatred of equality, are wanting in equity." He doubted his
writing could transform the prejudices of his fellow citizens, and again and again returned to the idea that the real
job at hand was to transform himself ("Not being able to
govern events, I govern myself") by questioning his own
judgment and in doing so possibly inspire others to do the
same. As he tweaked and appended his essays over the last
twenty years of his life, Montaigne was living out his belief that we only come to understand ourselves over time
and then never fully. He was less interested in drawing

conclusions about himself than having a conversation with himself. The dialogue would end only in death.

"To make a right judgment of a man, we are long and very observingly to follow his trace," wrote Montaigne, and I cherish this quote because it allows for contradiction and change. Whether by man, woman, or god I have no hope but to be judged on the balance of my actions over time, and I ain't the only one. In an essay comparing Montaigne to Kanye West, Chris Jackson wrote:

> Kanye is like Montaigne, who said of himself that he doesn't record being, but passing. That is, Kanye's raps aren't about a static, fixed identity as much as they are about the passing flow of thoughts through our consciousness, thoughts that are wild and contradictory and hard to justify in the light of day. They pulse with love and seconds later hate. Our thoughts are all over the place: they surge with unrealistic ego and then punish us with unrealistic doubt.

This passage sent me down a memory hole to seek out the book *I'll Take You There: Pop Music and the Urge for Transcendence* in which author Bill Friskics-Warren writes of how Johnny Cash strove for wholeness while navigating his own contradictions:

*He was a doubter and a believer, and he could be hip
as well as square, a rebel and a voice of reconciliation.
He was an addict and an evangelist, a protestor of the
war in Vietnam and a guest at the Nixon White House,
a singer of unexpurgated odes to murder like "Delia's
Gone" and an aficionado of clodhopper cornpone whose
second wife, June Carter Cash, was one of the funniest
comics in the history of country music.*

And yet the man whose first Number One hit was an
ode to fidelity written for the wife he would leave, was
simply following Montaigne's "long trace":

*In the end it was Cash's hard-won multiplicity, his
struggle—not nearly as facile as the admission, "I
find it very very easy to be true," in "I Walk the Line,"
claims—to remain true to his unruly heart that af-
forded him whatever measure of transcendence he
knew. And again, not by disavowing or collapsing the
tensions that dogged and defined him, but by embrac-
ing and lifting them up, just as he embraced and lifted
up people on society's margins and urged the rest of us
to open our hearts to them as well.*

That last line could serve as a summary of Montaigne
and his writing in his concluding years, when, according

to Donald Frame, he "urged simple human goodness" and "moderation rather than abstention or excess" and "a balance of duties between oneself and others."

I was raised on the Bible. You don't hit that final "Amen" in Revelation and close the book like it's over. So it is I read and reread Montaigne (and those who write about Montaigne),[54] forming a tenuous latticework of intersecting observations, ever aware that mine is an amateur exercise, but hoping to make some small progress toward the balance of my duties in this life. It is a process—like oiling my boots—to be repeated until I live out this line:

Death, it is said, releases from all our obligations.

This phrasing (from the Cohen translation of the *Essais*), with its use of the word *releases*, implicates us in our duty while we remain among the living. Montaigne may have retired to his castle, but he did not retire from every duty.[55] "When he contrasts the solidity of acts with

---

[54] Here's a Donald Frame quote best buried in a footnote of the final chapter of your book about Montaigne: "The best book about Montaigne was written long ago . . . by Montaigne himself."

[55] He did allow himself some wiggle room. When plague ravaged the countryside around him, he took his family and bailed; in some of his writing he revealed that while he deplored all injustice he did not always fight all injustice.

the futility of words," writes Starobinski, "he accepts the traditional moral teachings and opts for acts." Even after he began essaying he continued to serve as a soldier, an advisor, and the mayor of Bordeaux. Even as he lay dying he had agreed to travel to Paris to counsel the king in some affair of state. "The world is inapt to be cured," he wrote, but never forgot he stood on that world, and forsook it at his own peril. I read Montaigne in my room above the garage and think I better start speaking up more. That I should make a few more ambulance calls. Take up the cause of uncertainty, if nothing else. If reading and writing about Montaigne has taught me anything, it is not that I am on some path to perfection where I never again grab the pig fencer. Montaigne *is* the pig fencer, jolting me out of my absentminded musing and into the recognition that through the examination of my *im*perfections I can better serve my obligations to others.

This, above all, is what I take from Montaigne.

I am obligated.

I must do better.

ACKNOWLEDGMENTS

First and foremost, to my parents—anything decent is because of them; anything else is not their fault. Lisa and Berni for belief and paperwork. Jennifer, editing god and goad. The Harper crew that does so much for me even from a distance. Alissa no matter the time zone. Blakeley for adjustable boom stands. Dave for SneezingCow.com. Matt B. for math. Book boxers. Dan and Lisa and staff for cubes of quiet. Racy's and Mister Happy for the go-to corner. Nolte. Ben in Ohio. Colorado blended and extendeds. McDowell family. TFD and Emergicare for the fresh batteries in my pager.

Dr. Tressie McMillan Cottom, Daniel José Older, Zach Halmstad, Howard Bryant, and Justin Vernon for the early reads.

La Cheeserie!

The Long Beds, faithful musical road dogs, boon companions, and hearers of all my stories so many times you really oughta just hire *them*.

My family, mi familia.

Nobbern, where I am forever from.

My wife, who knows me better than me. My daughters, the same.

P.S. If I forgot you, please see chapter 3.

SOURCES (DIRECT AND INDIRECT)

BOOKS

The Art of the Personal Essay, anthology selected by Philip Lopate.

Bad Feminist, by Roxane Gay.

The Complete Essays of Montaigne, translated by Donald M. Frame.

Digital Sociologies, edited by Jesse Daniels, Karen Gregory, Tressie McMillan Cottom.

Essays, by Michel de Montaigne, translated with an introduction by J. M. Cohen.

Essays, by Michel de Montaigne, translated by John Florio, edited by Dr. John N. McArthur.

The Essays of Montaigne, Complete, by Michel de Montaigne, translated by Charles Cotton, edited by William Carew Hazlitt.

Examined Lives: From Socrates to Nietzsche, by James Miller.

Great Books, by David Denby.

How to Live: Or a Life of Montaigne in One Question and Twenty Attempts at an Answer, by Sarah Bakewell.

How to Read Montaigne, by Terence Cave.

I'll Take You There: Pop Music and the Urge for Transcendence, by Bill Friskics-Warren.

M, by Hannah Brooks-Motl.

Michel de Montaigne: The Complete Essays, translated and edited with an introduction and notes by M.A. Screech.

Montaigne, by Peter Burke.

Montaigne, by Hugo Friedrich.

Montaigne and Melancholy, by M.A. Screech.

Montaigne in Motion, by Jean Starobinski.

Montaigne's Discovery of Man: The Humanization of a Humanist, by Donald M. Frame.

Montaigne: Selected Essays, with La Boétie's Discourse on Voluntary Servitude, translated by James B. Atkinson and David Sices, with introduction and notes by James B. Atkinson.

Shakespeare's Montaigne: The Florio Translation of the Essays, A Selection, edited by Stephen Greenblatt and Peter G. Platt.

Some Renaissance Studies: Selected Articles 1951–1991 with a Bibliography, by M.A. Screech.

Standing by Words, by Wendell Berry.

The Swerve, by Stephen Greenblatt.

Triumph: The Power and the Glory of the Catholic Church, by H. W. Crocker III.

NEWSPAPERS, PERIODICALS, DIGITAL PUBLICATIONS, AND MISCELLANY

"12 Fundamentals of Writing 'The Other' (And The Self)," by Daniel José Older, January 15, 2014, BuzzFeed Books.

"Against the Normative World," April 9, 2015, interview of Ashon Crawley by Sofia Samatar on The New Inquiry website.

"Beyond Belief," post by John Horgan on his website.

Bryant, Howard, Twitter account: @hbryant42.

SOURCES (DIRECT AND INDIRECT)

"Confessions of an Aesthete," October 1, 2014, post by Terry Teachout on the *Commentary* blog.

"Cosmopolitanism as a Philosophy for Life in Our Time," by David Hansen, in Volume 14, 2013, of *Encounters in Theory and History of Education*.

"A Crash Course in Stoicism," by Donald Robertson, October 26, 2012, on the blog *Stoicism and the Art of Happiness*.

Heer, Jeet, Twitter account: @HeerJeet.

"Hip-Hop, Comedy, and the Great Kanye West Debate," by Chris Jackson, January 13, 2011, in *The Atlantic*.

"How Books Help Us to Be Better Human Beings," by Jonathan Bate, in the August 17, 2015, issue of *New Statesman*.

"Kidney Stone in My Shoe," by Sonya Huber, December 6, 2012, on the blog *Her Kind*.

"La Bella Vita," by John Armstrong, February 14, 2014, in the digital magazine *Aeon*.

McMillan Cottom, Tressie, Twitter account: @tressiemcphd.

"Michel de Montaigne, Time Traveler," by Hannah Brooks-Motl, February 29, 2016, on Literary Hub website.

"Montaigne on Self-Esteem," an episode of the Channel 4 television series *Philosophy: A Guide to Happiness*. Narrated by Alain de Botton.

"Montaigne: The Eclectic Pragmatist," by Anthony Long, Volume 1, Issue 2, *Republics of Letters*.

Montaigne Tweets (Twitter account) @TheDailyTry.

Older, Daniel José, Twitter account: @djolder.

The Partially Examined Life: A Philosophy Podcast and Philosophy Blog.

"Dr. Paul Rahe on the Hillsdale Dialogues Discussing the Work of Montaigne" October 6, 2014, radio interview with Hugh Hewitt.

"Pilgrim's Return," January 12, 2015, post by Terry Teachout on his blog *About Last Night*.

"The Sex We're Having Versus the Sex We Think We Should Be Having," August 4, 2015, post on *Broadly* by Jessa Crispin.

"The Shrink and the Sage," February 7, 2014, article by Antonia Macaro and Julian Baggini in FT Weekend (*Financial Times*).

So Many Books: The Agony and Ecstasy of a Reading Life, a blog containing numerous Montaigne-related posts by Stefanie Holl-michel.

"A Towering Intellect," book review by Colin Burrow in the November 11, 2003, edition of *The Guardian*.

"Traumatic Brain Injury and Therapeutic Creativity" by Michael Sperber in the April 16, 2012, issue of *Psychiatric Times*.

"Viva la Joia," by Roy Porter, in Volume 5, Number 24, December 22, 1983, issue of *London Review of Books*.

ABOUT THE AUTHOR

Michael Perry is a humorist, radio host, songwriter, playwright, and the *New York Times* bestselling author of several nonfiction books, including *Visiting Tom* and *Population 485*, and the novel *The Jesus Cow*. He lives in rural Wisconsin with his family and can be found online at SneezingCow.com.